Mr. Mohammad Seyedielmabad has more than 10 years of independent research experience on the current challenges of human life in the fields of water crisis, climate change and neurodegenerative diseases especially Alzheimer's with 10 papers published in Persian and English. His theory on the structure and function of human memory is a breakthrough in neuroscience that is the result of an innovative thinking. The theory has been nominated for the Think Tank Awards 2021 by Prospect Magazine of UK. It has also been nominated for the Early Career Investigator Award in 2021 by 10X Genomics and the Blas Frangione Award 2022 of USA. The book author has worked with world's leading publishers including IntechOpen and the IWA publishing as an author and reviewer. He is also in touch with other global publishers such as Frontiers. In addition, he has collaborated as a reviewer with several international journals and conferences.

I dedicate the book to my mother who taught me about morality, freedom and humanity, and to my father who taught me the faith, endeavour, continuity and courage.

Mohammad Seyedielmabad

The Meshk Technique for Alzheimer's Disease

Austin Macauley Publishers™
London • Cambridge • New York • Sharjah

Ordering Information
Quantity sales: Special discounts are available on quantity purchases by corporations, associations, and others. For details, contact the publisher at the address below.

Publisher's Cataloging-in-Publication data
Seyedielmabad, Mohammad
The Meshk Technique for Alzheimer's Disease

ISBN 9781685622725 (Paperback)
ISBN 9781685626242 (ePub e-book)

Library of Congress Control Number: 2023904316

www.austinmacauley.com/us

First Published 2023
Austin Macauley Publishers LLC
40 Wall Street, 33rd Floor, Suite 3302
New York, NY 10005
USA

mail-usa@austinmacauley.com
+1 (646) 5125767

This book is the result of a comprehensive study on the structure and function of memory in humans and Alzheimer's disease. The medical information in this book is not advice and should not be treated as such. Do not substitute this information for the medical advice of physicians. The information is general and intended to better inform readers of their health care. Always consult your doctor for your individual needs.

I would like to thank my sister, Ms. Homa Seyedielmabad, for her financial and spiritual support to complete and publish the book.

Table of Contents

About the Book

This book reviews the crucial role of memory in the human brain. For this purpose, previous investigations and researches about the construction and function of memory are studied. The mechanism of the memory function is reviewed, and crucial drivers for the working of the memory are indicated. Then an applied memory model that could serve as a framework to study the memory function is also introduced. Therefore, the memory unit is introduced as a basic information structure. Also, a structured platform for the memory unit is determined for encoding the information and data in the human brain. Then a pattern of information coding is detected. Thus, a basic framework to study the memory function is conceived. The results of this investigation pave the way for the discovery of a basic algorithm to understand the memory function in the human. Also, the relationship between the memory structure and Alzheimer's disease and also the mechanism of memory in the Alzheimer's disease is detected. Therefore, the book introduces a simple way to overcome Alzheimer disease (AD). This way can be applied to the prevention and treatment of AD. Furthermore, memory pyramid and its structure component in human brain is introduced.

Consequently, an applied and simple solution to overcome the Alzheimer's is introduced. This study could help to scientist to more investigation about neurodegenerative disorders in the human.

The Meshk technique for the prevention and treatment of Alzheimer's disease has been developed based on the principles of Meshk theory. The theory has been nominated for the Think Tank Awards 2021, in which Dr. Shah Nawaz Ali Mohamudally, Associate Professor of University of Technology in Mauritius as a validator, has given his opinion on it as following:

The nominee addressed the function of the human brain memory by investigating into its fundamental structure. Moreover, he developed a novel framework to model the memory function. The 'Meshk' model proposed is actually based on triple drivers, analogue to Einstein relativity theory. The research work broadens understanding about the brain memory functions from diverse perspectives. Furthermore, it adds original contributions to the body of knowledge regarding music therapy in the context of diseases linked to memory lost or disorders such as the Alzheimer disease.

Author Statement

The Meshk theory introduced by the author for the first time and published as a chapter in the book of 'New Frontiers in Brain-Computer Interface' on February 13, 2020. This theory provides a novel approach to study the construction and function of memory in human beings. Furthermore, the theory introduces an innovative solution to overcome the Alzheimer's disease (AD) which is a novel technique in this way. The present book is written based of the Meshk theory and the detail of its structure. The goal of publishing this book is reach to a better cognition of the human memory and mind in a simple and applied way for overcoming the brain and mind disorders.

The imagination is more important than knowledge.

Albert Einstein

Chapter 1
A New Classification of the Structure and Function of Human Memory

1.1 Introduction

Memory is one of the greatest unsolved mysteries to have faced the human race. There are a lot of questions about memory. Scientists all over the world have investigated this issue, but so far no definite results have been obtained. Only recently has it been determined that memory is a faculty of the mind. It is the medium by which information is encoded, stored, and retrieved in a region which is called 'hippocampus'. Also, memory is vital to experiences and related to the limbic system of brain. Models of memory provide abstract representations of how memory is believed to work. There are several models that are proposed over the years by various psychologists. Controversy is involved as to whether several memory structures exist[(3)]. For decades, neuroscientists have attempted to unravel how the brain makes memories[(17)]. Atkinson-Shiffrin and Working Memory are the different kinds of memory models that have been available in the last decades. The human brain is estimated to have approximately 86 billion neurons[11], and each neuron has tens of thousands of synapses[2], leading to over one hundred trillion synaptic connections. On top of this astronomical complexity, one needs to map each connection or neuron to a given stimulus, yet possible numbers of stimuli that can be used are infinite given the complex, ever-changing nature of the world we live in[20]. As such, the unifying mathematical principle upon which evolution constructs the brain's basic wiring and computational logic represent one of the top most difficult and unsolved meta-problems in neuroscience[1, 4]. Because of that, it is required to introduce an innovative approach to solve this problem. Recently, Dr. Tsien and his colleagues have taken an alternative 'thought experiment' approach to this problem[13]. It seems that this approach has an important contribution in the area of neuroscience. One useful concept in pursuing this line of reasoning is cell assembly, a term coined by Hebb (1949) to describe the supposed computational building block or computational primitive in the brain[6, 15]. This notion has attracted keen interest, especially with emerging large-scale recording techniques[12]. Hebbian cell assembly was postulated to be comprised of a group of neurons with strong excitatory connections that are formed after learning[12, 10]. They focus their research on the hippocampus, particularly a region called CA1, which is important to forming memories of events and places in both people and rodents. Their results add to a growing body of work indicating that a linear flow of signals from one neuron to another is not enough to explain how the brain represents perceptions and memories[14, 19]. Rather the coordinated activity of large populations of neurons is needed. According to the theory of connectivity by Dr. Tsien, a clique is a group of neurons that respond similarly to a select event and thus operate collectively as a robust coding unit(13). Table 1.1 indicates each event as three blocks are active and the others are inactive. The investigators represented clique activity as a string of binary codes that revealed details of the event an animal experienced(16). In the string fragments shown here, 1 means a particular clique is active, and 0 signifies inactivity(16).

A	1	1	0	0	1
B	1	1	0	1	0

Table 1.1 The binary codes, (A) earthquake (B) elevator drop

The idea of quantum coherent waves in the neuronal network is derived from Frohlich(81). He viewed these waves as a means by which order could be maintained in living systems and argued that the neuronal network could support the long-range correlation of dipoles(81). Repeated stimulation in a continuous and rhythmic way is a critical factor in operating the coupling process. As more and more memory blocks are turned on and active, more processes of recalling the internal information are performed. Therefore, more information and data are extracted from the fundamental memory. In the human brain, there are millions of memory blocks that are not used during the lifetime. Thus, people have entire blocks in fundamental memory but only turn on some of the blocks in their lifetime (Table 1.2). The active blocks make up the central memory in the human brain. The recalling of internal information and data from the fundamental memory is different about people. This process is different from elementary to advanced levels. It is dependent on the quality and the quantity function of coupling process in the memory.

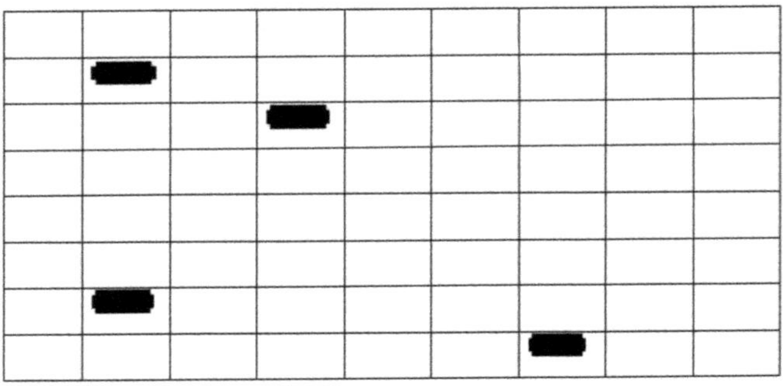

Table 1.2 The building blocks of fundamental memory in the human brain, black blocks are the memory in use and white blocks are the inactive memory

Alzheimer's disease (AD) is a global health issue given the increasing prevalence rate and the limitations of drug effects. Also, Alzheimer's disease (AD) is a progressive neurodegenerative disorder with characteristic clinical and pathological features(1). It is now recognized as the most common dementing disorder in the elderly(2). Management of these patients presents a multidisciplinary challenge(1, 2). Possible, but not proven, protective factors against AD include the use of estrogen replacement therapy, use of NSAIDs, education and challenging occupations(5). AD results in cortical atrophy which particularly affects the temporal lobes and hippocampus. The clinical presentation of the disease reflects the distribution of pathology in these brain areas and consequently, problems with higher cortical function are prominent. The clinical hallmark of dementia associated with AD is the gradual onset and progression of short-term memory loss and other cognitive functions[5, 8]. Almost invariably the first symptom is short term memory loss and forgetting appointments. As the AD progresses, the patient typically has difficulty performing routine tasks at home or work. Ultimately the patient becomes less able to perform basic activities of daily living and self-care functions[8]. Agitation, anxiety, restlessness, depression, day-night disorientation, delusions, hallucinations, and other behavioral disturbances frequently complicate the middle stages, of AD[8]. In the final stages, the patient may be unable to recognize family members and unable to communicate[8].

As a consequent, non-pharmacological interventions are of importance[53]. One of the important solutions to Alzheimer disease is the Music Therapy (MT) that scientists are investigating about. The music therapy is a non-pharmacological way with a long history of use and a fine usability for dementia patients[53]. Dr. Rong Fang and colleagues summarized different techniques, diverse clinical trials, and the mechanisms of MT as it is helpful to the cognition in AD, providing reference for future research[53]. Many articles have demonstrated that MT can reduce cognitive decline especially in autobiographical and episodic memories, psychomotor speed, executive function domains, and global cognition. MT is a promising intervention for strategy of dementia especially of AD and it must be started as early as possible(53). Kim HJ et al. demonstrated that multi-domain cognitive stimulation including music therapy could improve the word-list recognition and recall test scores. Besides, the domain of community affairs of AD patients and the Quality of Life (QOL)-AD of caregivers were also better(69).

1.2 Triple Drivers

In Einstein's special theory of relativity, $E = mc^2$, energy and mass are equivalent and transmutable. It is possible to explain the relativity theory of Einstein for the brain. Overall, the memory function in the human brain is dependent on three principles drivers (Figure 1.1). Each one of these drivers inherently provides one of the essences of the human brain as shown below:

1- Water equivalence conducted by the water driver ⟶ C^2
2- Nunch resonance provided by the energy driver ⟶ E
3- Substance (Mass) proportion provided by the mass driver ⟶ M

The human memory in the function circle is based on three basic drivers including water, energy, and substances. Water is a driver for creation and operation of memory. The Quantum Brain Dynamics (QBD) theory claims that water comprises 70% of the brain, and proposes that the electric dipoles of the water molecules constitute a quantum field[36]. Therefore, human memory works in a general circle provided by water called 'Water Equivalence (WE)'. While, memory neurons are working, WE – is an important factor for coding and encoding memory information. The last studies have shown that conscious experience correlates not with the number of neurons firing, but with the synchrony of that firing[39]. Ears are likely to be related to this process because the ear is connected to the brain's equilibrium process. For a long time, the oldest part of the brain was thought of as a 'control room' for human motions[40]. Now there is evidence that cerebellum also stores the temporal information about the music we are listening to and then it recalls this information when while reproducing the music[40]. Moreover, an amazing fact is that cerebellum was discovered to be a center for emotions[40]. Also, the water equivalence of brain can lead to a powerful electromagnetic field in the brain.

The electromagnetic theories of consciousness propose that consciousness can be understood as an electromagnetic phenomenon[37]. Electromagnetic field theories of consciousness propose that consciousness results when a brain produces an electromagnetic field with specific characteristics[37, 38]. There is a mechanism in the human brain which by 'Nunch Resonance (NR)' enables the brain to accelerate and increase reactions to stimuli from the environment. According to figure, each nunchaku includes two parts that are linked together by a connection. A part of the nunchaku inputs the force and energy and then transmits it to the other part by a connection. Owing to the resonance in this process, the output force is much more than the input. It is what McFadden termed as 'amplifying the microscopic quantum effects'. In the CEMI theory, the synchronous firing of neurons is argued to amplify the influence of the brain's electromagnetic field fluctuations to a much greater extent than would be possible with the unsynchronized firing of neurons[38]. The mineral substances in the function circle of the human brain provided by a process called to 'Substance or Mass Proportion (SP)'. It is likely that heart and sense organs are important parts of this process. The concurrence of the triple drivers is required for encoding of information in the memory. Triple driver compatibility is required to encoding information in memory. According to Einstein's special relativity theory, $E = mc^2$, energy and mass are equivalent and transmutable. In this equation, c^2 is the determinative factor, and so it can be concluded that (WE) is the main driver for recalling information in the brain.

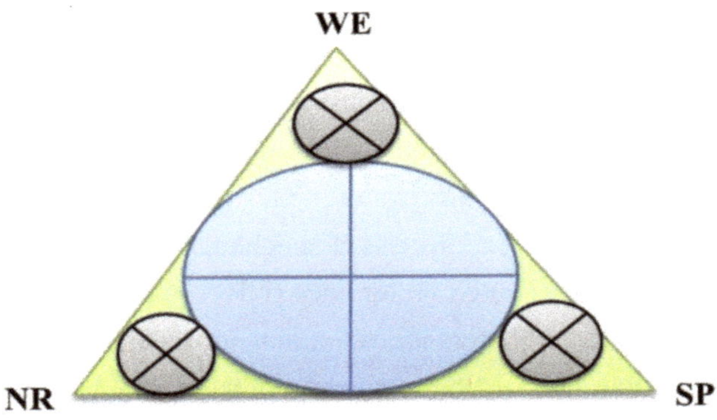

Figure 1.1 The diagram of triple drivers that inherently provide the essences of human brain, the gray, blue circles and green triangular indicate the NR, SP and WE

1.3 Triple Memory

Most of the memory models presented in the past decades have been based on time. The terms of memory, including short-term, long-term, and working memory, are defined by time. In Einstein's special theory of relativity, $E = mc^2$, time and space are not invariable. According to the Meshk theory(23), space and time are variable and parts of the inherent memory. The word Meshk is known in the Sumerian and Kurdish languages to the human brain. Therefore, memory in the human brain is indicated in a new model called 'Manna model'. In this new model, memory is divided into three parts fundamental, central, and peripheral as the fundamental memory is inherent intelligence. The base structure of memory in this model is shown in Figure 1.2.

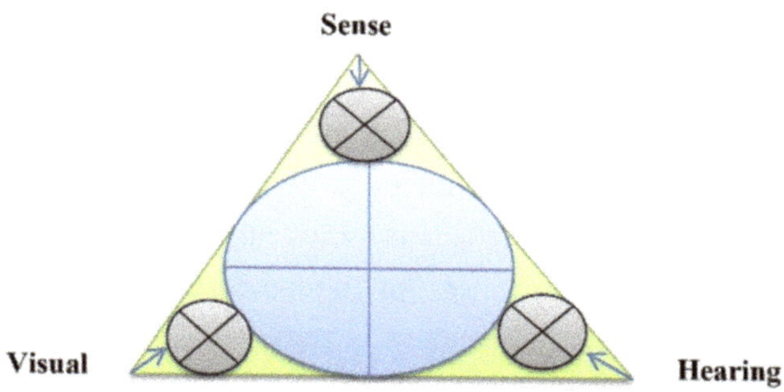

Figure 1.2 The diagram of Manna model, the gray circles, blue circle and green triangular indicate the peripheral, central and the fundamental memories

The Fundamental memory can be called intrinsic memory and has existed since the advent of mankind[(23)]. This memory is very cryptic and so unknown so far. The spatial and temporal information are parts of the memory nature. The ability of infants to swim and learn a language is clearly an example that confirms the nature of fundamental memory in human beings. All learning processes in the central memory of human are connected to fundamental memory that is called ‘recalling of the internal information’. The environmental information includes visual, auditory, and sensory data. The eye, ear, and heart are the centers of receiving the visual, auditory, and sensory information from the environment and transmitting them to the brain (Figure 1.2).

It is possible that sensitive organs send their signals to the heart and, after coordination, the information is sent from the heart to the brain. The peripheral memory is specialized to receive environmental information from the eye, ear, and heart and transmit them to the central memory. At the start, external information is transferred from the peripheral to the central memory. In the process of being transferred, some of the data is removed, and some remains. By coupling of blocks in the fundamental to the central memory, internal information is retrieved. The ‘Coupling Process’ is a turning point in reminding of internal information from the fundamental memory (Figure 1.3). The Water Equivalence (WE) is an important factor for operating the coupling process. The coupling process can be the starting point for neural momentum in the brain neural circuit. Therefore, in human memory, it is necessary to receive information from the auditory, visual, and sensory systems for making memory coding unit. At the same time, the information in the fundamental and central memory is also binded as binary codes. The coupling process leads to the creation of this important structure. This is what Dr. Tsien and colleagues termed as 'neural cliques'. They discovered that these overall network-level patterns are generated by distinct subsets of neural populations or neural cliques(16). The two strands of memory, akin to strands of Deoxyribonucleic acid (DNA), run in reciprocal connection and are thus parallel. The two strands of memory are coupled together for making a crucial structure. The base strand is made in the fundamental memory and the couple one is in the central memory. These two strands make Human Memory Strand (HMS) that enables the brain to advanced learning and behaviors in humans. In this structure, information processing is done in a simple way by the human brain. Therefore, it could possibly a fast way to distinguish problems and issues.

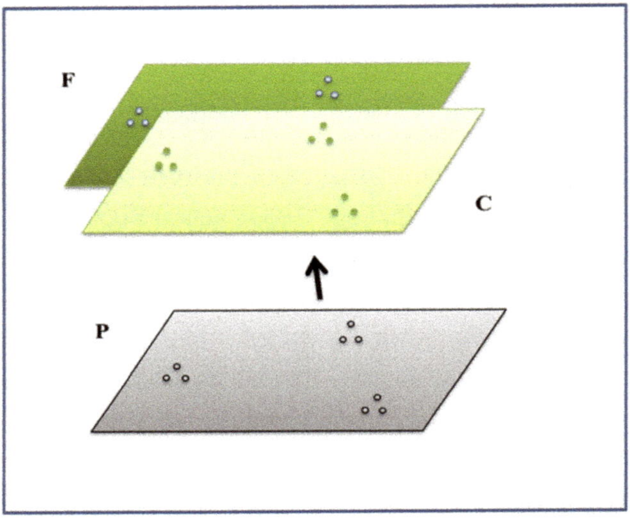

Figure 1.3 The coupling of memory blocks in the human brain. Plates F, C, and P show the fundamental, central and peripheral memories, respectively

There is whatever you want in the memory, the memory is a kind of pharmacy or chemistry laboratory, in which you sometimes find a soothing medicine and sometimes a dangerous poison.

Marcel Proust

Chapter 2
Human Memory Strand

2.1 Memory Structural Coding

The following sequence shows the function of memory strand.

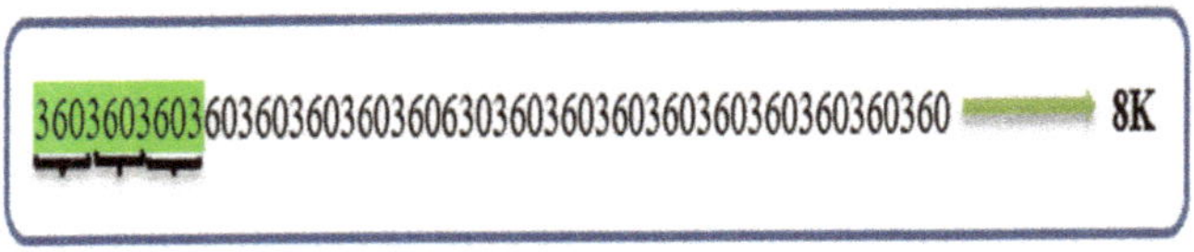

If the 3603 code is assumed to be repeated consecutively, it can also be represented as a sequence of 360 triple codes. The mechanism of creating a structural pattern of human memory by the triple codes is shown in Figure 2.1. This structural sequence contains important parts that can be called pauses or rest times. The numbers zero in this sequence indicate pause times. Dr. John Medina (2008) stated that the human infant brain does not look like a whiteboard because there is a default on it(24). He also showed that the human nervous system develops after birth(24). According to this, the developing process of human brain continue from the birth until to death. There are processes of learning, thinking, language, and the ability to recall and storing information in human creation. The fundamental memory is default of human brain, and peripheral and central memory develop throughout life. The central memory is the connection between fundamental and peripheral memory. In the peripheral memory, olfactory sense, tactile sensation and sense of taste are related to the sense (cardiovascular system), visual and hearing systems, respectively (Figure 2.2). The olfactory sense can be related to the heart system and therefore can affect other parts of the peripheral memory, including the vision and hearing systems, and thus affect the decision-making process. Brian Tracy mentioned in the book Focal Point that the eye is unable to see important things, but the heart can see the truth clearly[(78)].

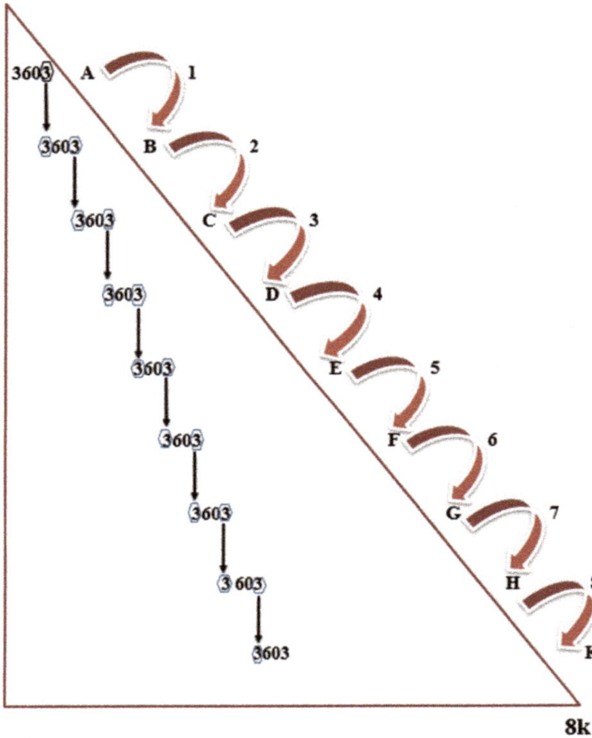

Figure 2.1 The memory structural template

One of the first symptoms of the last pandemic, Coronavirus COVID-19, in patients is the inability of olfactory sense to detect of odors. According to the relationship between the cardiovascular system and olfactory sense, the first symptoms indicate that the cardiovascular system can be affected by the virus. Also, the olfactory sense relates to Amygdala in the brain and has an important impact on memory. This evidence suggests that COVID-19 can impair human memory and lead to amnesia or Alzheimer's disease and other dementias. The tactile sensation in the blind people is very important for creating the mirror image and imagination therefore helps them to recognize the path. Also, the touch ability helps them to read and receive information in the learning process. The tongue is an important part of the gustatory system. It is also related to the auditory system because of its role in creating sound. If the auditory system in humans is destroyed, the conciseness is disturbed. The blind people can see things with tactile sensation and keep the texts in mind. They are also able to recall what they have heard before due to the unique ability to imagine.

In a study by researchers of Hebrew University in Jerusalem, tested the performance of 19 congenitally blind individuals and individually matched sighted controls in two types of memory tasks: item memory and serial memory(50). In the item-memory tasks, subjects were asked to identify 20 words from a list they heard. In the serial-memory tasks, subjects had to remember not only the words, but also their ordinal position in the list. Those who were blind recalled more words than the sighted, indicating a better memory overall, they found. Their greatest advantage, however, was the ability to remember longer word sequences according to their original order(50). The blind individuals' remarkable edge in item recall resulted not from a specific advantage in remembering the first words in the list, or the most recent words. Rather, the blind showed a better memory for all of the words, regardless of where they fell. That result suggested that the key to their success may lie in representing item lists as word chains, perhaps by generating associations between adjacent items[50].

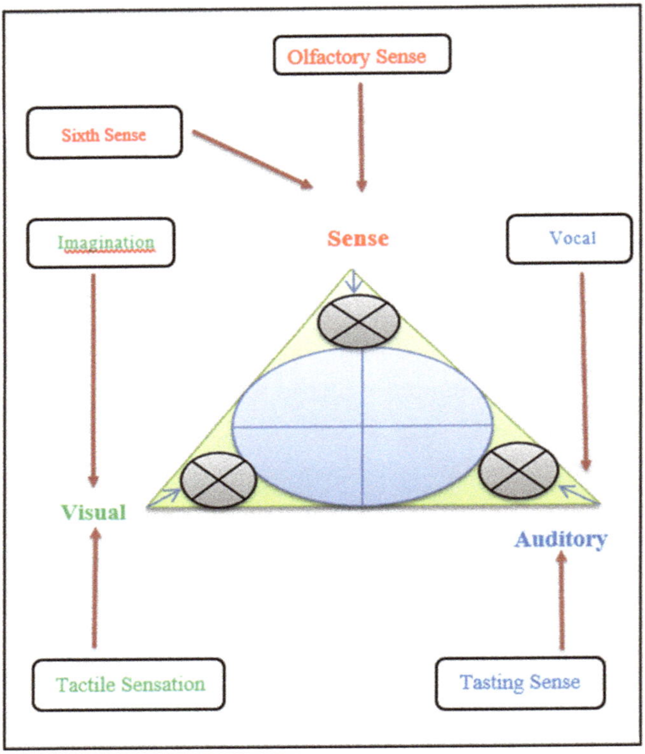

Figure 2.2 The relationship between the main and secondary senses

The results by Dr. Achille Pasqualotto, who led the study for the University of Bath, and fellow researcher Dr. Michael Proulx showed that congenitally blind participants reported significantly more correct words than both late-onset blind and sighted people[51]. In addition, the congenitally blind participants formed fewer false memories of what they had heard. Furthermore, they showed that visual experience has a significant negative impact on accuracy of semantic memory[51]. Consequently, research has shown that people often remember words related to those we actually hear[51].

2.2 Memory Unit

Music can simultaneously stimulate all parts of the peripheral memory system including sensory, visual and auditory sections. Also, memory stimulation by music can create a connection tunnel between the peripheral and fundamental memory for remembering information. This is a tunnel between the light and dark parts of memory that are connected to each other. This connection, coupling process, is created in the central memory and is an essential part of the memory function. There is a need for a paradigm shift from behaviorist stimulus-response concepts toward notions of predictive coding in self-organizing recurrent networks with high dimensional dynamic[42]. Neural networks consist of nerve cells that are linked by many reciprocal connections[43] and are capable of non-linear computations[44]. Neuronal networks with non-linear neurons and densely connected feedback loops can generate dynamics that is more complex, variable and rich than expected[45]. Therefore, the two structures of the hippocampus located in the limbic system can operate in reciprocal connection to the coding of the information in the brain. According to the Meshk theory, the information and data received from the environment encode in a structure is called 'Memory Unit'[23]. Each memory unit is actually a perception unit in the human brain. It is necessary to operate one or a few perception units to figure out the problems and issues. According to the Manna model features described in the previous sections, each memory unit includes three parts visual, auditory, and sensory[(23)]. Therefore, each perception unit makes three functional codes (Fig. 2.3). At the same time, the central and fundamental codes are binded together by coupling process to create the binary codes described in the previous section. For an individual event, three sections in the memory unit have to work and each section has a functional code. It is likely to operate numbers of the memory units for an individual event. This is dependent on the quality and quantity of internal information that might be extracted from the fundamental memory. Locating consciousness in the brain's Electromagnetic (EM) field, rather than the neurons, has the advantage of neatly accounting for how information located in millions of neurons scattered through the brain can be unified into a single conscious. In this way, EM field consciousness can be considered to be 'joined-up information'[(38)]. When neurons fire together, their EM fields generate stronger EM field disturbances[(38)]. Therefore, synchronous neuron firing will tend to have a larger impact on the brain's EM field (and thereby consciousness) than the firing of individual neurons[(38)]. However, their generation by synchronous firing is not the only important characteristic of conscious electromagnetic fields as in Pockett's original theory, spatial pattern is the defining feature of a conscious field[(37)]. In Meshk's theory, spatial and temporal patterns are part of the intrinsic memory features[(23)]. The CA1 region of the hippocampus receives inputs from many brain regions and sensory systems, and this feature most likely influences what type of information a given clique encodes[17, 18]. The nervous system can be seen as a nested hierarchy of nonlinear complex networks of molecules, cells, microcircuits, and brain regions[41].

2.3 Manna Alphabet

The Manna alphabet has a simple structure for understanding the brain. It is an appropriate programming system for the human brain and could be the first one designed by humans. In this way, the alphabet conforms to the structure of information coding in human memory. Therefore, in the coding process, signals received by the brain from the environment translate into special characters. These characters arrange based on the external information received from the environment. A recorder strand of information divides into four parts and each part includes a memory unit (three memory codes) (Fig. 2.4).

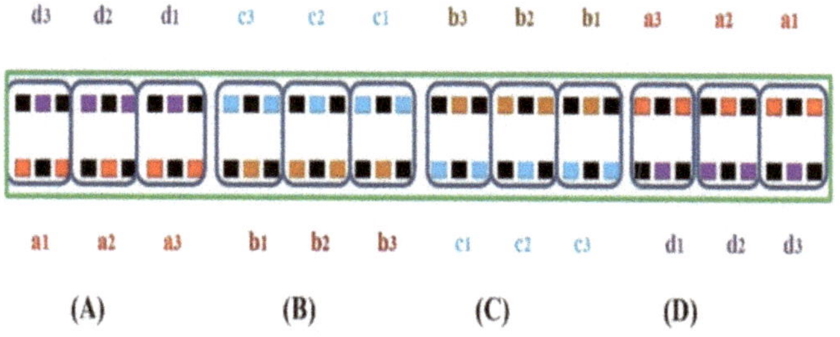

Figure 2.3 The pattern of human memory coding; A, B, C, and D – are the memory units

The Manna alphabet framework is interestingly conformed to the algorithm of the human brain structure (Table 2.1). This alphabet could be the best basic format to receive and transmit data and information by the human. Manna is a term that concerns people to the south of Urmia Lake in the east of Kurdistan[7]. There are a lot of mysteries about the Manna that has not been discovered until now. The only thing that has been distinguished is that the people of that age were blessed with great intelligence. They developed music, agriculture, animal husbandry, industry, medicine, and astronomy. According to ancient literature, Sumerians and Mannea had an advanced alphabet with 37 and 36 letters, respectively[9, 21]. The Mannea were an ancient ethnic group, and a rich trove of literature written by them exists[5]. This literature and books covered various topics including music, agriculture, astronomy, medicine, industry, and more importantly water engineering. The Manna alphabet included four partitions with a specific algorithm, and each part had nine letters. This algorithm made a basic structure for humans to communicate and understand the relation between the people from the ancient period until now. This alphabet can be a suitable pattern for decoding memory architecture in humans.

	1	2	3	4	5	6	7	8	9
A (a1, a2,a3)	a	i	u	ka	ku	xa	ga	gu	ca
B (b1,b2,b3)	ja	ji	da	di	du	ta	tu	Θa	ca
C (c1,c2,c3)	na	nu	pa	fa	ba	ma	mi	mu	ya
D (d1,d2,d3)	ra	ri	ia	va	vi	sa	Ŝa	za	ha

Table 2.1 The structure of Manna alphabet between 600 and 800 BC.

Bisotoun or Bisotun is a mount of the Zagros Mountain, which is located in eastern Kurdistan (Figure 2.4). It is well known for the famous Bisotun Inscription and rock relief from 500 BC. Mount Bisotun has the largest stone arch in the world, built since ancient times. Part of the Bisotun has a large carving on the rock in the shape of a lotus flower with nine petals (Figure 2.4). The number nine, the biggest single odd number, is regarded in ancient Chinese philosophy as the maximum yang number, representing the masculine or positive principle in nature, and since it is believed to be supreme it is therefore reserved for emperors only[(28)]. The horizontal and vertical rows of door knobs in the Forbidden City, the largest and best-preserved palace in the word, must be arranged according to the imperial hierarchy[(28)]. The only exception in this city is East Flowery Gate, which has 72 (8X9) knobs on it[(28)].

Figure 2.4 The lotus flower carved in a part of Bisotun mount in eastern Kurdistan

The information isn't actually lost but is somehow encoded in the radiation black holes emit.

Stephen Hawking

Chapter 3
Memory Features

3.1 Hippocampus Structure

The hippocampal structure has four sections, including CA1, CA2, CA3, and CA4, each of which can be divided into nine partitions (Figure 3.1).

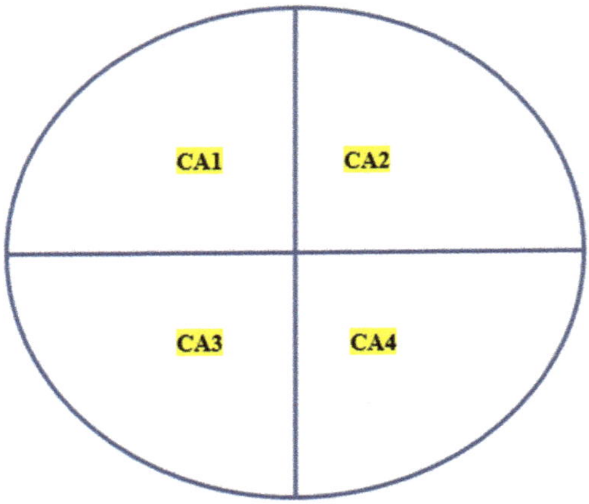

Figure 3.1 Hippocampal structure partitions

The structure of the hippocampus leads to a high speed of information processing (Figure 3.2). The hippocampal structure has three main features that control processing in memory:

1- Structure and shape (aerodynamic structure) → Speed
2- Balace of forces (crystal composition) → Balance
3- Resonance → Power

The structural shape of the hippocampus is similar to the wings of a fly. (Figure 3.3). The wing of a fly is an important part of its body. Flies have two wings and two halters to move and maintain balance during flight. Halters are advanced sensory-mechanical organs in flies.

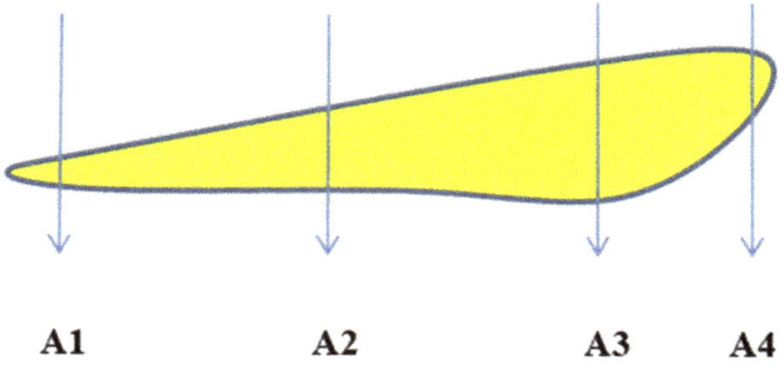

Figure 3.2 The hippocampal shape, like to the wings of a fly

The crystal composition and wing shape of flies lead to equilibrium at high speed in the air. This specialty greatly increases the maneuverability of flies. Flies have 100,000 neurons that control twelve muscles in his body because each muscle is controlled by a neuron to move the wings and create maneuverability and balance in the air. The housefly's thorax features all of its limbs used for movement. Though flies appear to have only one pair of wings, closer inspection reveals the presence of tiny, secondary wings, called halters, located below the main pair[(95)]. While their ancient ancestors had two complete sets of wings, houseflies are more efficient with fewer wings[(95)]. Over millions of years, the lower set evolved into the shrunken appendages you see today[(95)]. They're far from useless, however. Houseflies flap their halters at high speeds during flight, using them to maintain balance in the air[(95)]. If one halter is removed, the insect can only fly in circles. Without either halter, the fly can't take to the air at all[(95)].

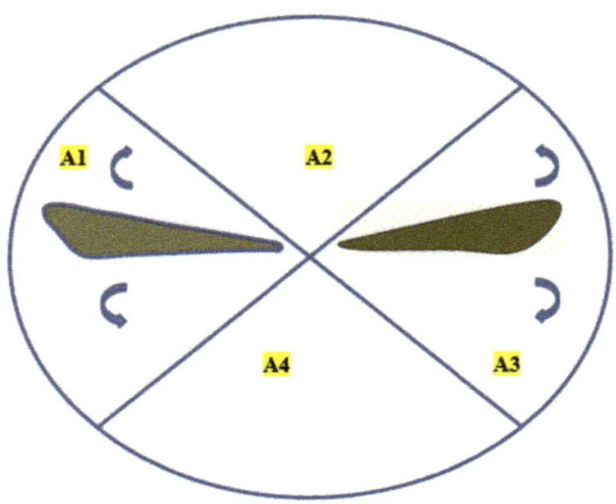

Figure 3.3 The aerodynamic structure of wing in flies

When flying a fly, three factors work to balance flight:

1- Wings……Tornado (Circle Front Edge) ⟶ Aerodynamic forces
2- Nervous system……Controlling part ⟶ Conducting
3- Muscles……power and speed muscles ⟶ Motor

In flight, the two wings of flies are in balance and can move in two different directions in the air. The wings aerodynamic of the fly is like to the aerodynamic shapes of two ski plates while a ski-man is moving in the high speed. The balance of a skier is created by the result of the forces on two ski plates (Figure 3.4).

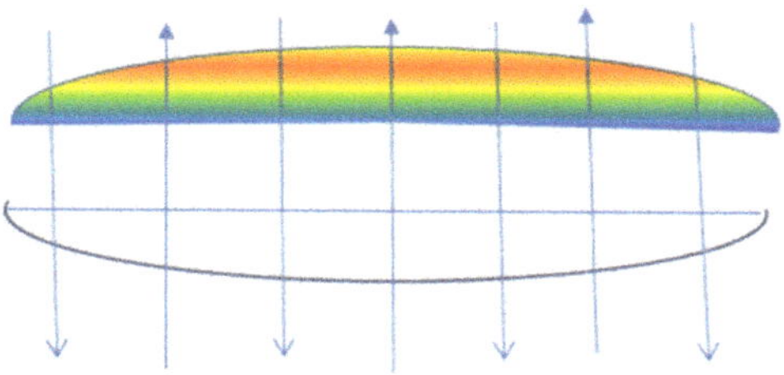

Figure 3.4 The result of forces on two ski boards in an equilibrium condition

Two hippocampus structure of human act with together to information processing in the high speed and accuracy. Without each part of the hippocampus structure, the information processing could be imperfect because it locates in out of the equivalence and normal situation. An experiment by Dr. Fang Wang and colleagues include a total of 36 unique visual cues that thirty-two cues were organized into eight A-B pairs and eight C-D pairs[32]. In this study, the four remaining cues were used to form four control cue pairs (E-E). Also, on each trial of preconditioning phase, participants (N= 24) were sequentially presented with pairs of cues.

(A ⟶ B, C ⟶ D, or E ⟶ E).

Participants were asked to respond with a button press if the second cue was different from the first[32]. The latency of these responds decreased linearly over the course of the preconditioning runs ($t\ [17] = -5.13$, $p = 4.2 \times 10^{-5}$), suggesting that participants acquired information about the predictive relationship between the cues[32]. They also analyzed medial and lateral Orbitofrontal Cortex (OFC) separately because their differential connectivity patterns suggest different functions. Specially, the lateral Orbitofrontal Cortex (OFC), receives input from sensory areas, whereas the medial OFC receives input from medial temporal lobe, including HPC, Hippocampus[32].

3.2 Memory Propulsion

The neural network in the human brain is made up of many working blocks. The components of each block are connected to create a specific task or signal. The levels of each block in the neural strand are p, q and t which are interconnected (Figure 3.5). The structure of each block and also the connecting way of blocks to create the neural stream indicated in Figure 3.6 A and B, respectively. Also, the dimensionless equation of neural stream in the human brain can be shown in Table 3.1. The fundamental memory is actually the subconscious mind. Therefore, according to this approach, the peripheral memory is the conscious mind and the connection of the subconscious mind and conscious mind occurs in the central memory space.

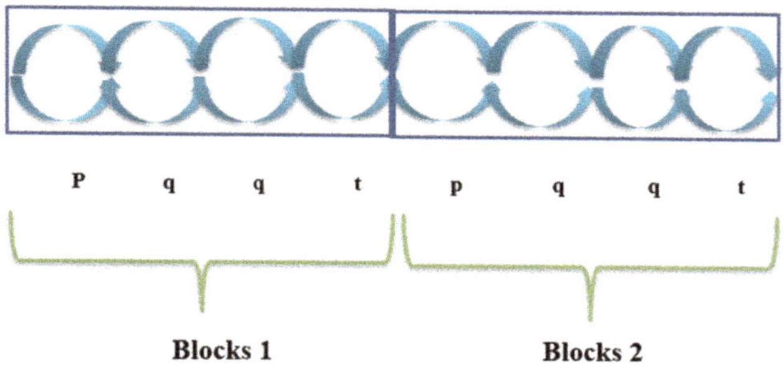

Figure 3.5 The diagram of levels of each block in the neural strand

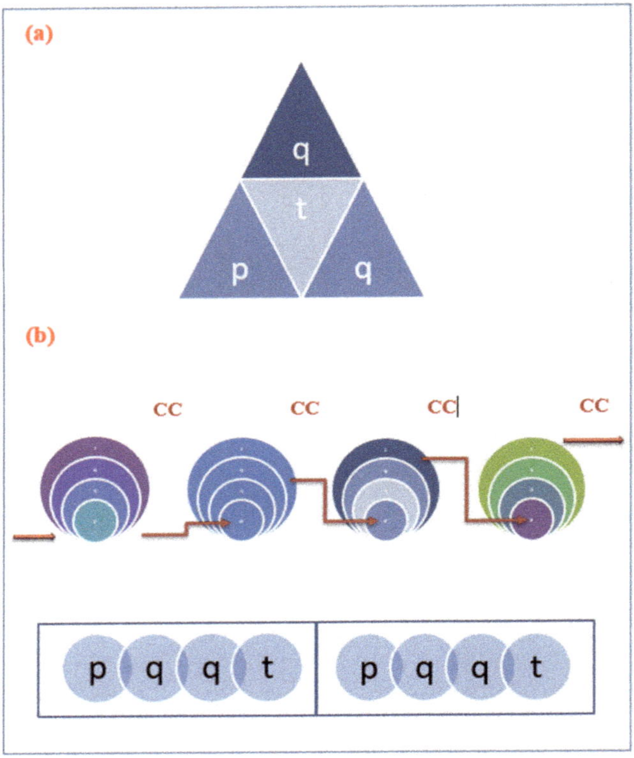

Figure 3.6 a. The structure of each block, b. The diagram of transitive levels, Concave Collision (CC), in the neural strand

In the past, scientists believed that the human infant's brain was like a whiteboard with nothing written on it[(24)]. Today, there is evidence that the infant brain has a basic information structure. The subconscious mind controls all aspects of human intelligence and therefore human life, unlike other beings, depends on the subconscious mind or fundamental memory[(25, 26)]. The learning process, which is a kind of recall of information from fundamental memory, consists of two parts, descending and ascending. In the descending part, the external information enters the memory and then this information comes down. This means that at this level, part of information deleted and part remains in the peripheral memory. In the next step, the information ascends during the coupling process as information extracted from the fundamental memory leads to the ascend performance. Liking to the herbs that have numerous kernels, the neural strand blocks play the role of kernels of memory (Figure 3.7).

Figure 3.7 The kernels of a bell pepper, similar to the human brain

These memory kernels remain to store the information in the brain. Also, the connection point between the blocks in this structure, p and t, can be called 'Concave Collision (CC)' or collision of two curves. As shown in Figure 3.8 B, concave collisions are the steps of the ladder in the nervous system because the neural strand are organized in these spaces.

$8K = (n)\, pq^2 t \longrightarrow K = (n)\, pq^2 t / 8$ (Eq. 1)
K: *latent power of kernels in the neural stream*
n: *number of memory units*
p: *point five, power of neural momentum*
q: *quarter levels, value of stream velocity*
t: *terminal point - power of emission*

Table 3.1 The dimensionless equation of the kernels latent power in the neural stream

According to Figure 3.8 A, the green circle is stable and the blue circles are rotating, causing pause or delay in movement. Rotating 9 times by rotating circles and then pausing by a stable circle creates a wave. Q is the point of change in neural stream. This point shows the time when the movement potential becomes action (Figure 3.8 B).

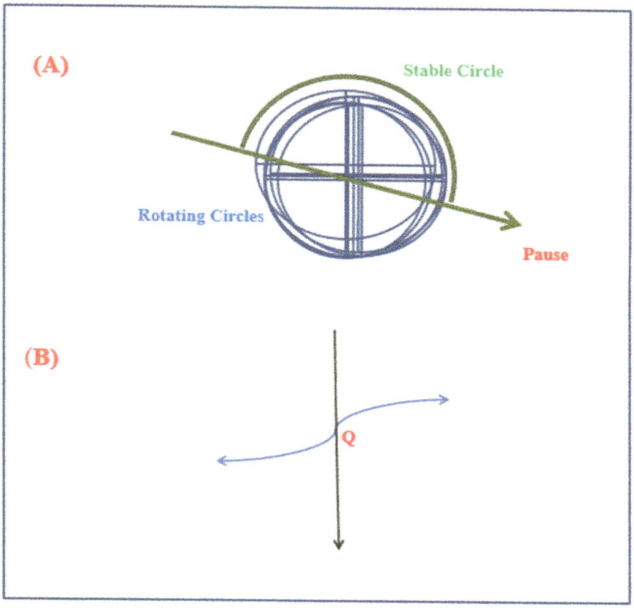

Figure 3.8 A. The creation of a neural wave, B. Q is a point of change in neural stream

According to the experiments of Dr. Libet (1985), there is a time lag between the initiate of neural momentum and the decision to take an action, known as the 'Libet delay'.[86]. In the Meshk theory, this time delay is called 'rest time' which is the time to convert of the potential phase into action, and is equal to 350 milliseconds in the neural circuits. According to Dr. Libet's trials, Ready Potential (RP) happens before action or reaction begins[86, 87]. While the action stage and the final processing of information in the neural circuit is underway, there is a 200-millisecond delay in muscles movement. Therefore, the time interval between the neural momentum and moving muscle is 0.5 second.

3.3 Memory Pyramid

A pyramid is a structure whose outer surfaces are triangular, making the shape roughly a pyramid in the geometric sense[98, 99]. The first pyramid-shaped structures were built in Mesopotamia, which are known as Ziggurat[98]. The Mesopotamians built the earliest pyramidal structures, called ziggurats[149]. Ziggurats were built by the Sumerians, Babylonians, Elamites, Akkadians, and Assyrians for local religions. The latest Mesopotamian ziggurats date from the sixth century BC[99]. If the base memory structure is considered to be a pyramid or as a 'memory pyramid', it can be shown in Figure 3.9. Accordingly, memory is divided into sections in which each section contains three structures that are memory codes. The base of the memory pyramid is the memory codes that make up the memory units. Therefore, each side of the pyramid is a memory partition, consisting of six memory codes or two memory unites to keep the pyramid in a solid and permanent condition. Therefore, the memory units A, B, C and D can be seen in the structure of the memory pyramid, which forms one of two helices of the memory strand. A complete pyramid contains all partitions of a memory strand or a double helix. This means that both sides of the pyramid are reciprocally coupled.

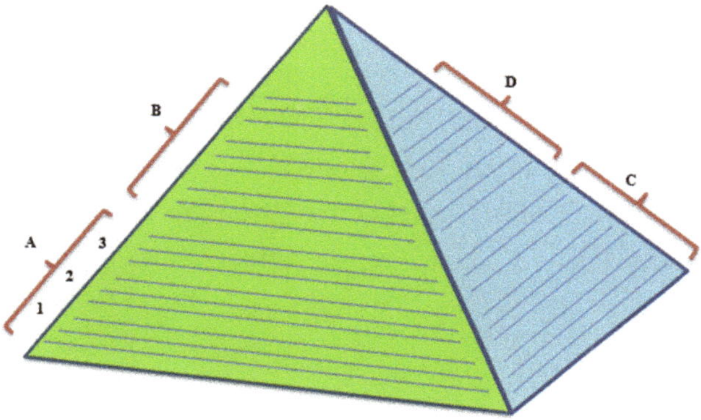

Figure 3.9 The structure of a memory pyramid; A, B, C, and D – are memory unites in the structure

3.4 Brain Musical Stream

A comprehensive study of human memory is needed to introduce a simple and practical way to overcome Alzheimer's disease (AD). This can be the invention of a new technique in the field of music therapy for Alzheimer's disease. Music therapy can be a simple and inexpensive procedure that takes less time to prevent and treat AD. According to Dr. Mehrabian, there are three elements to our interest in someone who conveys a message about their feelings: words are 7%, voice is 38%, and body language is 55%(22). To communicate effectively and meaningfully about emotions, these three parts of the message must support each other – they must be 'congruent'. In case of any incongruence, the receiver of message might be confused by messages coming from in inconsistent directions[22]. This suggests that the transmission of neural messages is similar to the transmission of music in which harmony is important. The basics of musical stream transmitted by nerve cells to generate a functional code or memory code as well as the neuro-musical stream pattern are shown in Figures 3.10 and 3.11, respectively. In the neural circuit of the human brain, there are three factors that support each other to create effective and meaningful stream.

Figure 3.10 The base of neuro-musical stream

Music is known as a part of mathematics and composers work in the field of mathematical rules. In mathematics, it has a 360-degree circle, so all mathematical rules are in the 360-degree range. Therefore, human memory encodes information in a complete circle including basic and complementary directions. The base direction (circle A), complementary direction (circle B) and the full direction (circle C) of memory coding are shown in Figure 3.12.

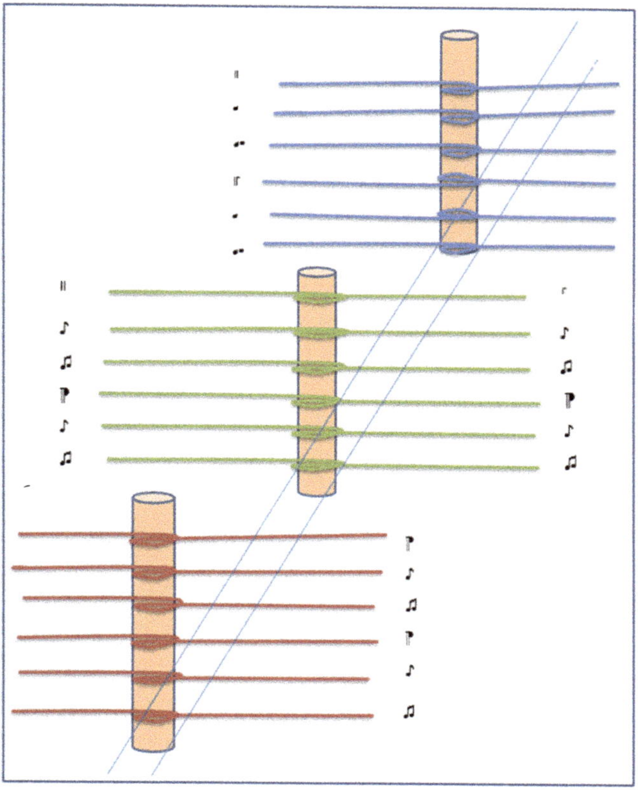

Figure 3.11 The neural stream pattern in a musical form

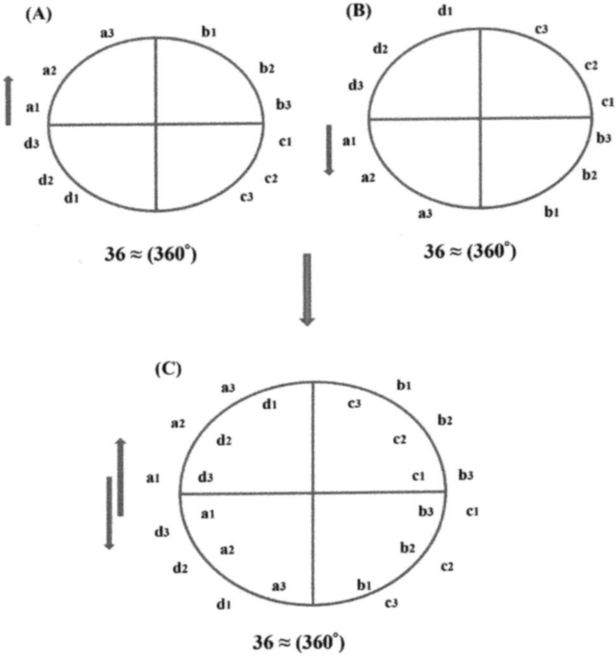

Figure 3.12 The circles that indicate the memory coding directions, (A) base direction of the memory coding, (B) complementary direction of the memory coding and (C) entire directions of the memory coding

Barbad is one of the first ancient musicians in Kurdistan who created 360 types of music by 200 AC. Kurdistan is a strategic region that was divided into four parts between Iran, Iraq, Turkey and Syria after World War I[75, 76]. Kurd is a term that concerns people in Kurdistan who are related to the ancient Sumerians[9]. Sumerians are known to the first founder of human civilization in the word. Kurdistan is a historical place that had governments in the ancient periods. One of them is known as Sassanian. In this period, music was prevalent, and many people worked on mathematics, and so, music was built on the structure and function of the human brain. According to this approach, music shows the inherence of human memory. Hence, music can relieve and improve the nerves and memory. The last studies that have been done by a lot of scientists prove this claim. Some of these investigations led to the wonderful discovery.

3.5 Dark Memory

The structural model of how nerve transmission is carried out is shown in Figure 3.13. In this pattern, the main part of this structure named 'Point 5' is located in its center. In the other word, point 5 is the generation center of signal in the human brain. All signals in the human brain are released from this point. When a stone is thrown into the water, waves are created at the place where the stone meets the water and spread around. Like this, in neural circuits, point 5 is the source point. Point 5 also plays the role of a black hole in the human brain. These points have properties similar to black holes in space. Absorption of water, energy and materials and then the emergence of waves from these points can create the structure of memory. The Meshk theory could answer the question of Stephen William Hawking's paradox of information. Also, the introduction of fundamental memory in this theory can solve the problem of Super-Translation in Hawking theory[(29)].

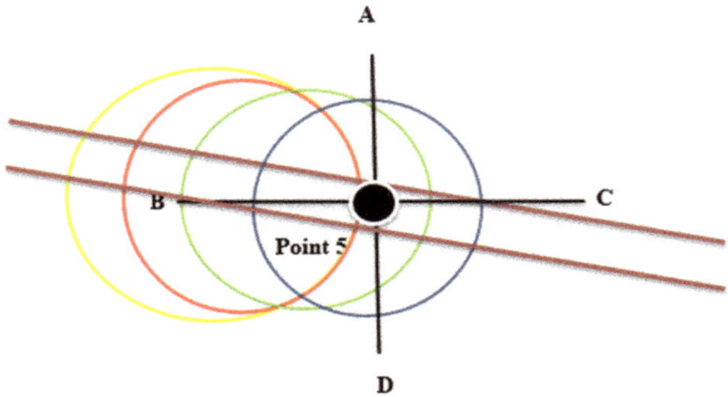

Figure 3.13 The template of neurotransmission; A, B, C and D – are memory partitions; Point 5 acts as a source of nerve signals

In the early 2000s, Hawking proposed that the information isn't actually lost but is somehow encoded in the radiation black holes emit. But neither Hawking nor his colleagues found a reason for what they called the information paradox[(80)]. Hawking believed that black holes might serve as gateways to parallel universes and that missing information might actually be transferred to those universes[(80)].

3.6 Spheral Memory

The pattern of the eyeball and brain for creating and sending waves is shown in Figure 3.14. The type of wave expansion is also shown in this figure. Contraction, expansion, and interval times in memory lead to the beating and propagation of waves in the brain. Accordingly, expansion begins at Point 5 and continues in all directions in the brain. Durable and coordinated contraction and expansion in memory maintains brain function. Also, contraction and expansion causes the memory string to warp. In the world, space-time warping is one of the most important factors in maintaining balance and proportion. Like the flight of a bird that balances in the sky by contracting, expanding, and stabilizing of body and riding on air waves, memory maintains its ratio in the neural circle. The ripples emitted from Point 5 are affected by external information received from the peripheral memory. Exciting recent research shows that non-trivial quantum effects are present in biological systems and not just in spite of, but sometimes because of, the interaction with the noisy and warm environment(41). Furthermore, because the brain is a complex non-linear system with high sensitivity to small fluctuations, it is likely that it can amplify microscopic quantum effects(41). The light photons that enter the eye also play a role in this process. In General Relativity (GR) and the equivalence principle developed by Einstein, it is said that black holes are area of space where gravitational attraction is very strong, because even light cannot escape. It is possible that there are spots with the characteristic of black energy in the frontal lobe of the human brain.

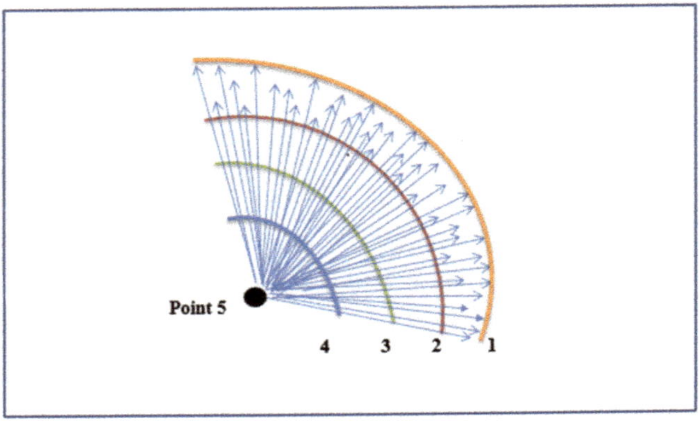

Figure 3.14 The functional similarity of eyeball and brain

The structure of the eye and memory is like a sphere consisting of five partitions (Figure 3.15). Point 5 exists in two general modes of action and latency. At the latency stage, the memory is in a potential state, and at the operational level, the memory is working. In the state of health, this cycle is repeated regularly. Unlike healthy conditions, in Alzheimer's disease these stages are irregular. The connection of two forces at point 5 launch the memory. Like seeds, memory needs a period of dormancy to grow. In Alzheimer's disease, dormant moments are either absent or incomplete. This problem leads to disorder in the point 5 and therefore causes memory impairment. Point 5 is similar to the retina in the eye and can be called the 'brain retina'. Also, the structure of the galaxy is similar to the structure of the eye and the brain, where black holes are galactic retina with the same features. This feature is well seen in the Potbelly Hill in the north of Kurdistan.

Potbelly Hill (Central Hill) is the oldest site of civilization in the world ever discovered[31]. This archaeological site is located in the South-eastern Anatolia Region near of the city of Sanliurfa[27].

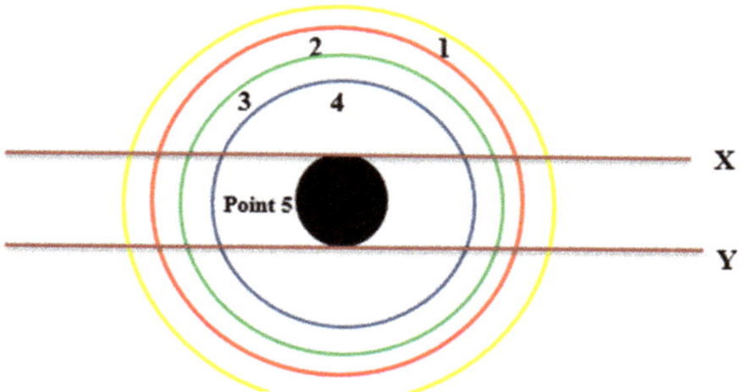

Figure 3.15 The eyeball and brain partitions

The eyes do not able to see the important things but heart can clearly see the rights

Brian Tracy

Chapter 4
Memory Booster and Transmitting Forces

4.1 Memory Transmitting Force

A water turbine converts the mechanical energy of water into electricity. Just like this, there are turbines in the human brain that use environmental energy to generate memory strand and ultimately intelligence (Figure 4.1). Also, the fundamental memory plays a major role in the production of human intelligence because it is the driving force of the human brain. This force controls and directs all the currents in the human brain because it is an inherent source. This can be called intrinsic memory and has existed since the advent of mankind, therefore this memory is very cryptic and so, unknown so far[(23)]. The human memory cycle is shown in Figure 4.1.The spatial and temporal information are parts of the memory nature[(23)]. Thus, space, time, and memory are intertwined to form a structure called Space-Time-Memory (STM).

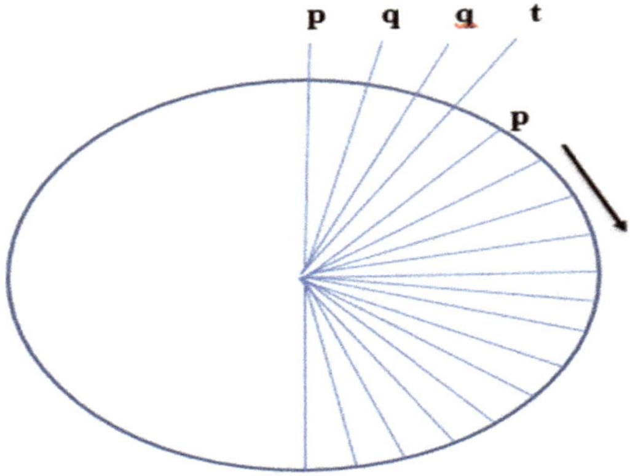

Figure 4.1 The functional cycle of the human memory

There is an information transmission axis between the fundamental and central memory that acts like the power transmission axis in a hydraulic machine or motor so that the coupling process plays exactly this role (Figure 4.2). By the coupling of blocks in the fundamental to the central memory, internal information is retrieved[(23)]. Coupling process is a turning point in reminding of internal information from the fundamental memory. The coupling process can be the starting point for neural momentum in the brain neural circuit. Therefore, in human memory, it is necessary to receive information from the auditory, visual and sensory systems to build a structural unit of memory.

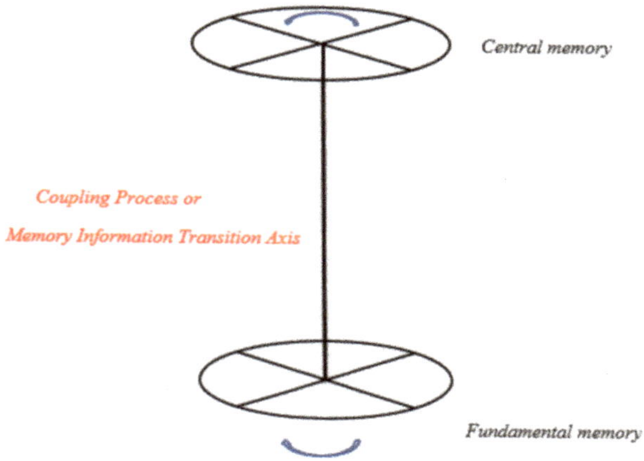

Figure 4.2 The axis of information transfer of human memory

4.2 Water Crystals and Memory Cores

Memory cores can be made of water crystals as these nuclei make up the entire neural network. These cores store memory, which leads to the recalling of information. In Alzheimer's disease, these nuclei are inactive and do not work properly. In plants, grains must have a dormant period to grow in the next growing season. If the seed does not have a dormant period, it will not grow. In human memory, a period of sleep is required for the functioning of memory nuclei. Sleep mediates between consciousness and death and acts as a dormant period of memory. In sleep and subconscious conditions, peripheral memory is least active because the sensory, visual, and auditory systems in the body are at rest. Therefore, the Water Equivalence (WE) and Substance Proportion (SP) are in equilibrium and the ear can hear sounds. According to the SP, the heart is in a state of relaxation, but it sends powerful messages to the brain as much as a state of full consciousness. Therefore, increasing Nunch resonance (NR) can severely stimulate fundamental memory and thus information recalling. This condition has many names in different sciences such as intuition, inspiration, creativity or meditation. In this process, many times more information is extracted from the fundamental memory than normal conditions. In the conscious electromagnetic information (CEMI) theory, McFadden proposes that the digital information from neurons is integrated to form a conscious electromagnetic information field in the brain(38). Since the brain is a 300° Kelvin tissue strongly associated with its environment(36), it is possible that the noise, compact and warm environment of the brain lead to transform the water crystals(23). This transformation, due to the inner equality of the brain, leads to the germination of innate memory. This process can form the basis of memory structure as what Dr. Masaru Emoto and colleagues (1996) studied about water crystals(22, 23). They proposed that human consciousness has an effect on the molecular structure of water and therefore emotional energies and vibrations could change the physical structure of water. Water memory is the purported ability of water to retain a memory of substances previously dissolved in it even after an arbitrary number of serial dilutions. It has been claimed to be a mechanism by which homeopathic remedies(35) work, even when they are diluted to the point that no molecule of the original substance remains. The brain activities in human infants have shown evidences the existence of the intrinsic memory(23). The high ability to learn language of infants is an important evidence that proves this claim. There is a fundamental potential in infants that enables them to advance learning and behaviors. This is a fundamental difference between human beings and others[(23)].

Astrocytes are star-shaped glial cells with arm-like extensions, and they weave among neurons in the brain[(52)]. They are support cells, and act in concert with neurons in many nervous system functions. A form of glucose called glycogen is stored inside these spider-like cells. As glycogen breaks down, it releases a substance called lactate, which gets transported by special proteins into neurons[(52)]. Scientists tested the importance of both the glycogen breakdown and the transport process to long-term memory function, and found that they both play a critical role in the ability to retain memories. Dr. Paul Aisen and colleagues proposed that astrocytes can be a source of proteins that support neurons, and there's substantial crosstalk between astrocytes and neurons but the details are difficult to study and not well understood[(52)]. They said that Alzheimer's disease is a top priority in this research but the research could also inform other memory impairment disorders, aging processes, and even diabetes[(52)].

4.3 Geometric Structure of Memory Strand

The geometric structure of the memory strand in humans is shown in Figure 4.3. This structure shows the different directions of memory for recalling information. While the learning process is recalling of information from the fundamental memory, it is required to make a connection between the fundamental and central memory(23). This connection is made by the coupling process. This process creates memory units and, therefore, leads to the formation of a memory coding strand. This structure is consistent with the structure and function of the hippocampus described in the previous chapter. Indeed, in the Manna model, the memory unit is a unit for recalling information. In this model, the coupling process is a turning point in the neural circuit and can be the basis of the memory function(23). Also, like the structure of deoxyribonucleic acid (DNA), memory is formed as a helical structure.

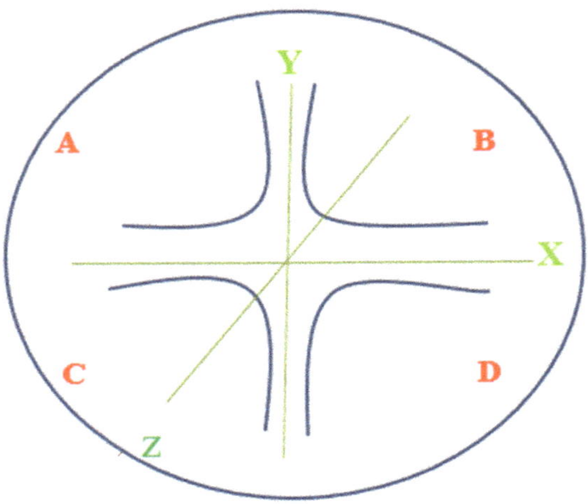

Figure 4.3 The geometric structure of memory strand

4.4 Memory Booster

Pahlevani and zourkhaneh rituals is the name inscribed by UNESCO for ancient sport. It is a traditional system of athletics originally used to train warriors in ancient times in Kurdistan[88]. In this sport, two sticks are used for exercises. Also, physical strength and flexibility are very important in this sport. In addition, the performance of ancient music and song plays an important role, so this sport has a great impact on the nervous system. As a result, ancient sports athletes have high memory and intelligence. Badminton is another sport in which the shape of the ball and the type of rotation of the ball are interesting. Figure 4.4 A shows the movement and rotation of a spiral of a badminton ball, which is like two snakes that twist together when listening to music and show how memory works in neural processes.

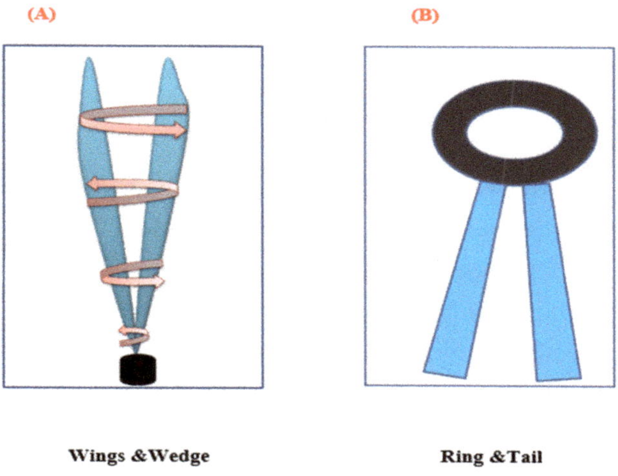

Figure 4.4 A. The badminton ball, B. The structure of a comet

The badminton ball is also similar to a comet that has two tail including gas and dust tail that move in a high speed (Figure 4.4 B). A comet is an icy, small solar system body that, when passing close to the sun, warms and begins to release gases, a process called outgassing. This produces a visible atmosphere or coma, and sometimes also a tail[(34)]. Furthermore, it is similar to Turban, a type of headwear, which Kurdish people have worn on the head from the ancient period. Kurdish dance is a symbol of solar system in the universe and also the human brain system (Figure 4.5). The memory structure has two part including dark and light sections that the fundamental memory is in the dark section and the peripheral memory is in the light one. The central memory locates between the dark and light sections as a connecting part.

The Kurdish dance is a circular group dance that Kurdish people have performed from the ancient periods until now. It is a form of a circle dance, with a single or a couple of figure dancers often added to the geometrical center of the dancing circle[(96)]. The dancers rotate in a circular course in two directions and perform rhythmic and systematic movements. In every dance, one dancer falls or comes to the head of the circle who is called 'Serchopi', holding a colorful or symbolic object in his/her right hand[(96)]. It is a tradition that no one takes his/her place until he/she leads the dancer's group at least one circle[(96)]. The rest of dancers are called 'Gawani' as sometimes is also primarily called to the last dancer of the circle[(96)]. The Serchopi and Gawani spin a strand object in their hands. The double strand objects could be a symbol of helix strand of DNA and also two strands of human memory that are connected together. The Kurdish music along with Kurdish dance have powerful impact on human intelligence and mind (Figure 4.5). Kurdish language, music and dance are three symbols of the historical memory of the people from ancient times to the present. This has been the best and most effective way to remember and preserve the language and culture of Kurdistan for thousands of years. For centuries, thousands of books and manuscripts by Kurdish scholars have been burned and destroyed by genocide in the war[97], but the Kurdish language, music, and dance have survived for hundreds of centuries. In addition, the magical power of Kurdish music and dance leads to the preservation of science, myth, art and other achievements of Kurdistan from generation to generation. Kurdistan is the oldest human settlement due to having the oldest Neanderthal dating from around 65,000–35,000 years ago in the world that has just been discovered. Shanidar Cave is an archaeological site located on Bradost Mountain in the Erbil Governorate of Kurdistan Region[30].

Figure 4.5 Kurdish dance, a symbol of the solar system and human intelligence

My method is different. I do not rush into actual work. When I get a new idea, I start at once building it up in imagination, and make improvements and operate the device in my mind. When I have gone so far as to embody everything in my invention, every possible improvement I can think of, and when I see no fault anywhere, I put into concrete from the final product of my brain.

Nikola Tesla

Chapter 5
Fundamental Memory Stimulation Techniques and Brain Honeycomb Network

5.1 Space-Time and Memory (STM)

According to general relativity of Einstein, the observed gravitational attraction between masse results from the warping of space and time by those masses. Human memory is made of U-shape components that wrap around each other. The curvature of the U-shape components increases efficiency in the memory structure and function. The two memory strands are interconnected to create a functional and unique architecture. In humans, this architecture works for advanced learning and behaviors. In Einstein's theory of relativity, space and time are represented as space-time structures. Transformation is one of the main features of this structure from the past to the future. The shape of the transformation can be represented as a curve that starts at a point, moves to a peak, and then falls. The third part of this structure is memory, which is woven into space and time and creates an invincible structure called Space-Time-Memory (STM) (Figure 5.1).

The type, amount, and shape of the space-time evolution are stored in a special memory called 'Memory Plate' and therefore remain (Figure 5.1). The memory plate is an essential part of the STM structure. While a particle or partition is stored in the STM, it remains so this structure is stable and invincible. Movement in the space-time-memory system is possible from two directions, one direction can be towards the future and the other towards the past. In human movement into space-time-memory, water is very important because water is the main factor of universal memory. It is likely that in the process of moving into space-time-memory, the human body does not lose its water and therefore does not need to drink water in this process because thirst does not occur. Under these conditions, due to water equivalence (WE), it is possible that the human brain is in a stable state and does not change during the movement into space-time-memory.

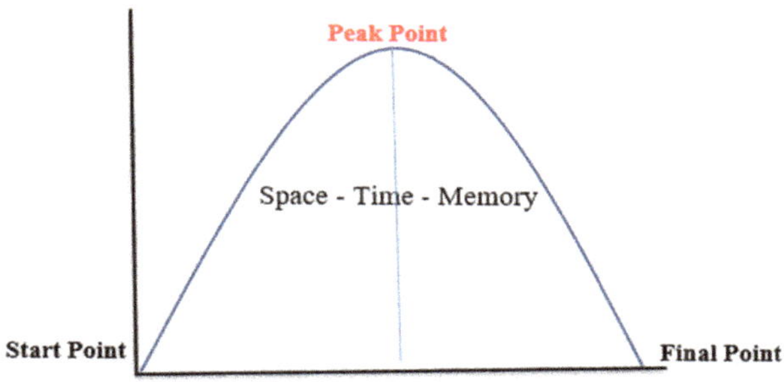

Figure 5.1 The stick points of space time memory (STM) curve

Space-Time-Memory (STM) Equivalence

- Space
- Memory
- Time

STM Equivalence Ratio =

$$3/100 = 0.03$$

The wormhole is said to be a speculation structure that connects different points in space-time and is based on a special solution of the Einstein field equations. The STM system is similar to the spider net, which has flexibility and conversion features.

5.2 Fundamental Memory Stimulation

5.2.1 Mirror Therapy Stimulation

Recently, doctors have been using mirror therapy to treat some limb pain. This method affects the minds of patients to achieve health. The principle of mirror therapy is the use of a mirror to create a reflective illusion of an affected limb in order to trick the brain into thinking movement has occurred without pain. It involves placing the affected limb behind a mirror, which is sited so the reflection of the opposing limb appears in place of the hidden limb(46). Mirror therapy was invented by V.S.

Ramachandran to help alleviate Phantom limb pain, in which patients feel they still have a pain in the limb even after having it amputated. Ramachandran first devised the technique in an attempt to help those with phantom limb pain resolve what they termed a 'learned paralysis' of the painful phantom limb[47, 48]. The visual feedback, from viewing the reflection of the intact limb in place of the phantom limb, made it possible for the patient to perceive movement in the phantom limb. Their hypothesis was that every time the patient attempted to move the paralyzed limb, they received sensory feedback (through vision and Proprioception) that the limb did not move. This feedback stamped itself into the brain circuitry through a process of Hebbian learning, so that, even when the limb was no longer present, the brain had learned that the limb (and subsequent phantom) was paralyzed. This system is thought to use the observation of movement to stimulate the motor processes which would be involved in that movement (See Triple Memory in Chapter 1). In fact, mirror therapy stimulates peripheral memory to effect on fundamental memory, which is essential for remembering information. In this process, stimulating the visual part of the peripheral memory is a key factor.

Satoh and colleagues indicated that listening to music and singing could improve psychomotor speed in Alzheimer's disease after 6-month treatment[60]. Decreasing Neuropsychiatric Inventory (NPI) score and prolonging sleep time were also observed after music therapy[60]. The clinical trial conducted by Arroyo-Anllo and colleagues showed that listening to unfamiliar music decreased the scores of MMSE and FAS, whereas the cognitive test scores did not vary in the familiar music group[62]. This indicates the preventive and protective effects of music therapy in the process of Alzheimer's disease. The above trials were mostly about mild or moderate dementia patients. Narme and colleagues compared MT with cooking therapy in moderate or severe AD or mixed dementia patients. Result showed that both music and cooking improved patients' emotions and ameliorated their behavioral disorders, yet no benefit was on the cognitive status[70]. In addition, Simmons-Stern NR et al. found that compared with spoken lyrics, AD and healthy older adults showed better scores on the memory test of sung general lyric content. This indicated that music could enhance the preferential sensibility to a familiarity-based content rather than increase the recollection of memory process[73]. Palisson and colleagues also verified that sung texts were better remembered than spoken texts for both AD and healthy groups, which indicated music during the encoding stage could facilitate learning and retention[64]. Nevertheless, another research conducted by Simmons-Stern NR and colleagues in 2010 showed that just AD patients had better recognition accuracy for the song lyrics than the spoken ones, which did not exist in healthy older adults' group[63]. So scientists proposed that music allowed a more holistic encoding to facilitate recognition only in AD patients, and music heightened arousal through better attention in AD patients[62]. In music therapy, especially in familiar music, the recall of information from fundamental memory is done by stimulating of all parts of the peripheral memory. This is a key difference between music therapy and other therapies because stimulation of all parts of the peripheral memory is essential in the treatment of AD patients.

5.2.2 Memory Photographical Stimulation

In photography, a dark space is needed to process the photos. Likewise, this process takes place in the human brain because fundamental memory must appear in a dark space to create intelligence. The fundamental memory plays the role of a photographic film or the original latent image in human intelligence. It is like an image that appears on the water upside down, so that in neuroscience this image is known as a 'Mirror Image' (Figure 5.2). The image processing takes place in the central memory and can appear in the real world as thinking, ideas, imagination or so on. Due to its effect on memory, music can stimulate the mind to create a mirror image that leads to the recollection of memories. On the other hand, conceptual processes such as inspiration, attraction rules, induction and treatment of patients without drug therapy or surgery could be explained by the recalling information from the fundamental to central memory. Also, speculative concepts such as mental wormholes are explained in this way. In the photographic processing, Silver Nitrate ($Ag^{+}NO3^{-}$) is used to produce the image. In the human mind, the process of imagination happened in the same way. It is possible that the presence of silver nanoparticles – AgNP – along with some substances can stimulate fundamental memory to recall information.

The study by Dr. Daniel Gonzalez-Carter and colleagues (2017) proposed that AgNP-detoxifying mechanism in microglia involving the sulfiding of Ag^+ ions and up-regulation of H_2S-synthesizing enzymes(33). This study demonstrated for the first time a correlation between AgNP-detoxifying mechanisms and reduction in microglia inflammation which serves a neuroprotective role(33). Also, a similar AgNP treatment of N27 dopaminergic neurons induced significant neurotoxicity, supporting silver ion sulfidation as a detoxifying mechanism inside microglia cells(33).

Figure 5.2 The mirror image of two climbers

5.2.3 Yoga Stimulation

Yoga is a group of physical, mental, and spiritual practices or disciplines which originated in ancient India. There is a technique known to Super Brain Yoga that the great Rishis of India developed this technique to increase the intelligence of people, using the principle of ear acupuncture points (Figure 5.3)(79). This technique is an ancient yogic practice that was designed to activate, energize and strengthen the brain cells[79].

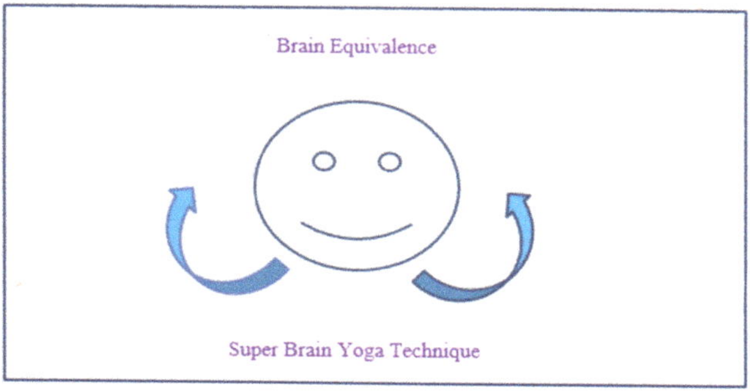

Figure 5.3 The Super Brain Yoga technique

According to the Meshk theory, this method can increase the brain's ability to store and process information because of its effect on the Water Equivalent (WE) driver (Figure 5.3) This method affects brain equivalence and results in water equivalent circle alignment.

5.3 Neural Honeycomb Network

The shape of the whole neural system is like a honeycomb network. In a beehive, the work is divided into many parts, each performed by a group of bees. Like a honeycomb, the neural network is made up of many workgroups that all parts in each group communicate together to create a specific task or signal (Figure 5.4).

Figure 5.4 The honeycomb neural network of human brain

Memory has a hierarchical structure so that in this structure some groups come together to form a large group. The memory structure is made up of small parts, memory units, that make up a complex system. This is the best practical form of memory performance. As a result, this is the simplest and most efficient way to do complex tasks. Also, the hierarchical system provides the best way to connect with other systems and the surrounding environment(43). In addition, if an error occurs in this system, the problem can be solved in the best way and as soon as possible. The general architecture and function of human memory in the Manna model is shown in Figure 5.5. Based on this approach, recalling internal information to form the Human Memory Strand (HMS) is carried out in five general stages, where A and B are potential levels and D and E are active levels in this process. In general, in the details of this model, eight levels of human memory function have been deciphered. Also, the basic algorithm for human memory is an exponential function. This is a transcendental function that shows the encoding characteristics of human memory as follows:

$$N = e^{i} - 1 \qquad \text{(Eq. 2)}$$

N is the number of memory units connected in different possible ways; e is the Napier's constant; i is the information they are receiving, and 1 is just part of the math that enables you to account for all possibilities.

As stated in the Mana Alphabet section in Chapter 2, the Mana alphabet framework is interestingly consistent with the human brain structure algorithm. The Manna language is close to the ancestral Sumerian language. The earliest language written in Mesopotamia was Sumerian, an agglutinative language isolate. Mesopotamia is a historical region in the west of Kurdistan situated within the Tigris–Euphrates River system[(74)]. Since World War I, Kurdistan has been divided between Iran, Iraq, Syria and Turkey[(76, 77)]. One of the important stories of Kurdistan is the epic of Gilgamesh, which is related to Mesopotamia[(74)]. The other old legend in Kurdistan is Shamaran myth, means to king of snakes, which is a mythical creature, half woman and half snake. The first human she encounters is a young man named Tamasp[(92)]. This is a Kurdish name that consists of the following two parts:

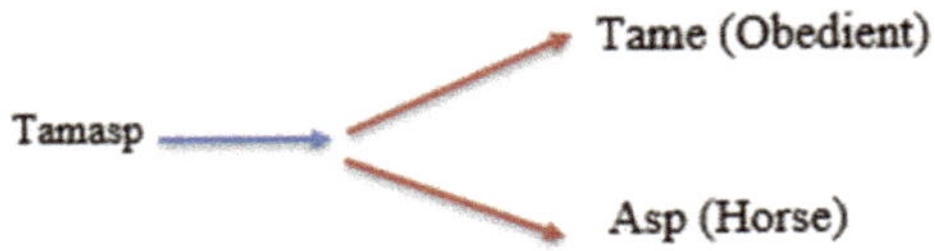

Horses and poultry (such as chickens and roosters) have been tamed for the first time in Kurdistan(91). The roosters have a special action, when the rooster wants to show interest in mating with the chicken, it rotates around the chicken in two different directions. He repeats this process over and over again so that the chicken remembers it. The official history of Kurdistan starts with the beginning of the Median Empire. The last king of the Medes is called Astyages, who, like the other kings of the Medes, ruled justly. In Persian texts, he has been introduced as Zahak, which is the Arabic word for Astyages, and many attempts have been made to destroy his character over hundreds of years. The most important source that Persian writings have used to introduce Astyages has been Ferdowsi's Shahnameh, which is not a historical book, but a book of poetry along with fiction and myth that is devoid of truth. In Persian texts, it is said that two snakes grew on Astyages shoulders, which is just a myth and not true. Contrary to what is mentioned in these writings, the snake is a sign of health and still a twisted snake is a symbol of the World Health Organization (WHO). Zoroastrian clerics have been pioneers of medical science in different periods of Kurdistan's history. They have used snake venom and oil to treat some diseases. In addition, two snakes coiled around each other symbolize the DNA molecule and the Human Memory Strand (HMS). There is still a traditional culture in Kurdistan of using snake oil and venom for medical purposes.

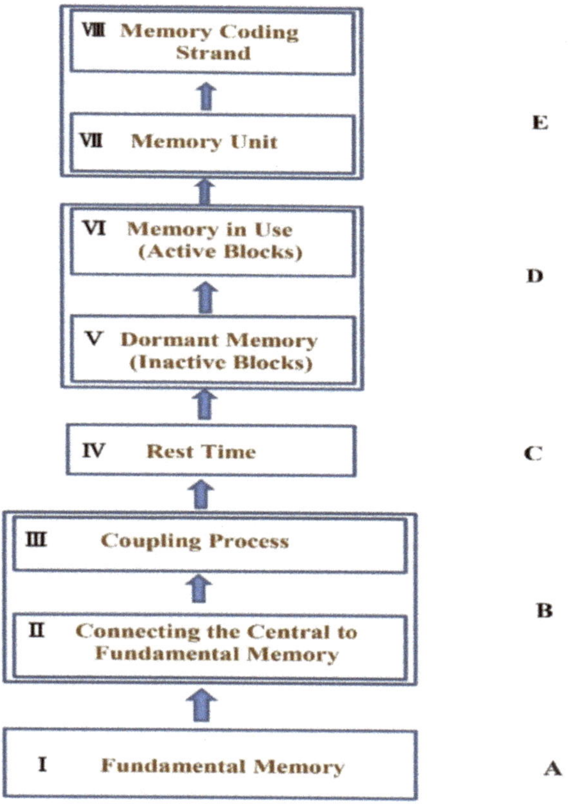

Figure 5.5 The general levels of Manna model in the Meshk Theory

(Storm-tide)
The tide said to the fisherman:
There are many reasons
why my waves are in a rage.
The most important is
that I am for the freedom of the fish
and against
the net

Sherko Bekes

Chapter 6
The Meshk Technique for Alzheimer's Disease

The results of the present study show that memory function in humans is influenced by the structure that consists of memory units (Figure 6.1). Memory units that have structural codes create the Human Memory Strand (HMS). In Alzheimer's disease (AD), one or more structural codes do not function properly (Figure 6.1). In this condition, irregularity in coding in part of the memory strand disrupts the entire of the system. Similar to train tracks, a malfunction in one unit causes a malfunction in the entire system. It is necessary to recover defective areas to prevent or treat the disorder. These sections are restored by initializing the correct structural codes in the memory encoding strand. In this process, the memory strand is repaired and thus the neural stream is adjusted in the correct way. Like the turbulence caused by water waves in the sea, the interference of brain waves creates turbulence in the memory, which leads to a disturbance in the brain water equivalent and Alzheimer's.

Alzheimer's disease, the most common type of dementia, is a neurodegenerative disease characterized by progressive cognitive impairment and neuropsychiatric symptoms which it causes patients to lose their daily living abilities(54). The pathological features of AD are accumulated amyloid beta (Aβ) protein in senile plaques and tau protein in neurofibrillary tangles, as well as the loss of neuron and synapse(57). AD is one of the greatest health challenges in this century that human encounter to it because along with the increasing of aging population, the numbers of AD and other dementias grow dramatically(57). Although Alzheimer's disease is more common in the elderly, new investigations show that the age at which people are diagnosed with Alzheimer's is decreasing significantly. This problem makes it necessary to find a basic solution for this progressive disease. There is a report which states the number of AD patients has reached over 35 million in worldwide from 2013, and this number is estimated to be triple by 2050(55). This report was made before the epidemic of Covid-19, and if it consider the effects of this epidemic in the exponential increase of Alzheimer's disease, this number will increase several times. There is lack of effective disease modifying drugs for AD(56), so considering the growing population of AD and the severe harmfulness to patients' health, non-pharmacological treatments play a more and more important role in preventing and relieving symptoms of AD, such as physical exercise, Music Therapy (MT), diet and so on(58). According to the structure and function of human memory that was stated in the previous chapters, music therapy plays an essential role in the prevention and treatment of this disease.

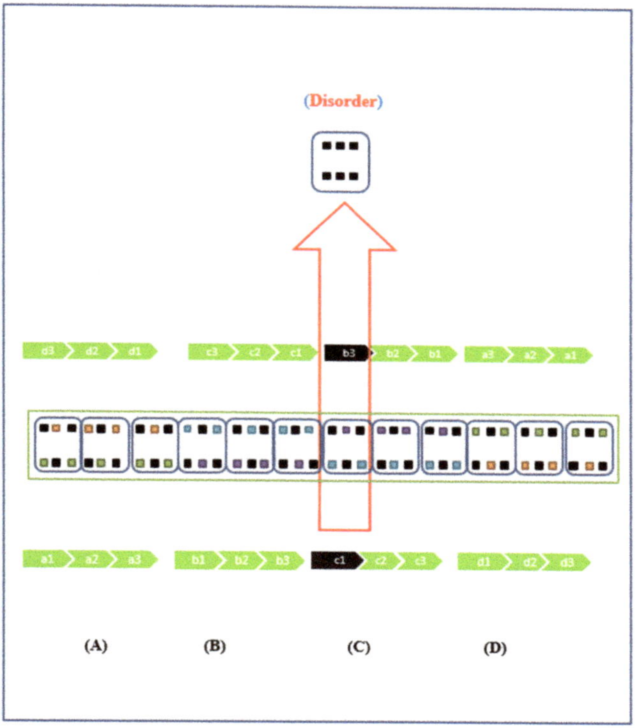

Figure 6.1 The diagram of disorder in the human memory coding; A, B, C, and D – are memory units

It is estimated that more than 50 million people live with Alzheimer's disease (AD) today[(100)]. Music Therapy (MT) is an important method for neuropsychological, cognitive, and social behavior goals in the field of dementia with low-cost[(59)]. It requires that research, practice, education, and clinical training are based on professional standards in MT

[World Federation of Music Therapy (WFMT)][60]. Many researches have demonstrated that MT is beneficial to improve cognition and to reduce neuropsychiatric syndromes of AD[60]. Due to the lack of side effects and the convenience to operate for AD patients and their caregivers, MT is an ideal intervention option. Dr. Rong Fang and colleagues proposed that more evidences with prospective, randomized, blinded, uniform and rigorous methodological investigations are needed. They also concluded that combining MT with other cognitive stimuli such as dance, physical exercise, video game and art should be considered. In the ancient culture of Kurdistan emphasized to learning of three mind-physical exercise including the horse riding, archery and swimming. At the same time as these exercises have an effect on the body, they also have a significant effect on the memory. The horseback riding along with archery and swimming can stimulate all parts of peripheral memory, including hearing, vision and sense, respectively (Figure 6.2). Therefore, the horse riding is a mind-balance exercise that effects the Water Equivalence (WE) driver in the human brain. The four horse gait rhythms including walk, trot, canter or lope and gallop are kinds of main balance rhythms that horse use to control and maintain of the movement. Also, the archery and swimming are the mind-physical exercises that effect on the Nunch Resonance (NR) and Substance Proportion (SP) drivers in the brain, respectively (Figure 6.2). As shown in Figure 3.14 (the spheral memory section of Chapter 3) throwing arrow from the bow, in the archery, is similar to propagation of waives from the point 5 in the memory. In human memory, thousands of these propagations occur in a neural momentum. These exercises are three innate human skills that are not forgotten over time and remain in the mind to be able to restore memory to its original state. One of the important points is that doing these skills properly leads to the coordination of neuromuscular system. The neuromuscular synchronization is a key factor in the prevention and treatment of dementia, especially Alzheimer's. It is also important to note that people can easily learn these skills from childhood to old age. There are many articles which have found that music therapy can reduce the mood symptoms and behavior disorders in dementias, especially depression, anxiety and agitation(66). Recently, scientists have paid more attention on the cognitive effect in AD patients with MT. Increasing articles have demonstrated that MT can improve multiple domains of cognitions in AD patients, including attention, psychomotor speed, memory, orientation and executive functions(71, 72). Bruer and other scientists found that listening to the music could increase the global cognition of AD(67). Ozdemir pointed out the effect of MT for AD could last for at least 3 weeks after intervention(68). After 6 weeks of intervention, Gómez Gallego and colleagues found that listening to the music which patients like could significantly improve the memory and orientation of AD(65). At the same time, improvements were observed in depression and anxiety in AD patients. In addition, anxiety was reduced in mild ones, and delirium, hallucinations, agitation, irritability, and language disorders were reduced in moderate AD ones(17). Katie Overy and colleagues (2009) proposed that the ability of music is to communicate social and affective information and to create the feeling of 'being together' that makes it so appealing to humans across all ages and cultures. They hypothesized that imitation, synchronization, and shared experience are all powerful components of successful therapeutic and educational music intervention activities[85]. Also, they suggest that people begin to conceive of music not only as 'humanly organized sound' and 'soundly organized humanity'[84], but also as shared affective motion experience, minimized prediction error, and as an extraordinary case of being together in time[85]. Melissa Leggieri and colleagues suggested that music techniques can be utilized in various ways to improve behavior and cognition[100]. They suggested that individualized music regimens provided the best outcomes for the patient. Music therapy can be considered a non-pharmacological intervention which has the potential effects to reduce cognitive decline, improve neuropsychiatric symptoms, and enhance the QOL of AD[61, 69, 70, 71]. Some researchers suggested that a treatment strategy should be developed individually based on the preferences and physical endurance of each patient[53].

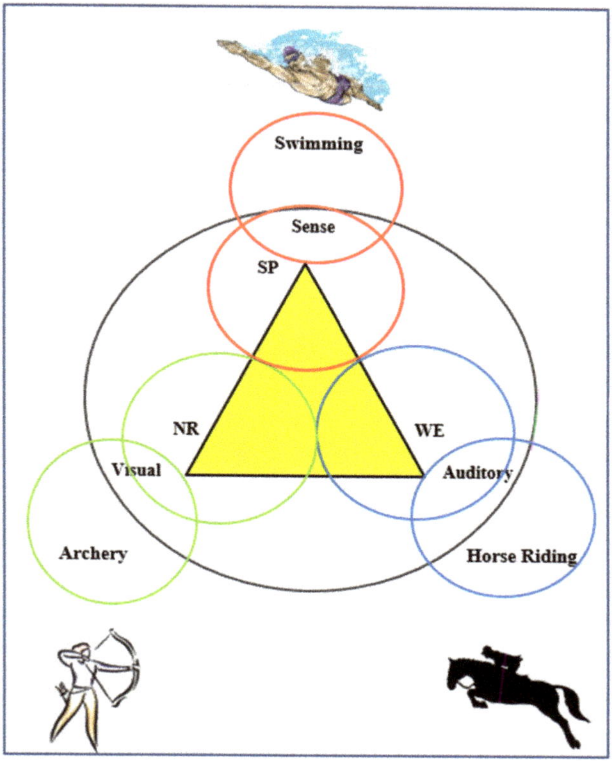

Figure 6.2 The impact of three ancient exercises on the peripheral memory cycle

The music utilizes a large variety of basic brain functions; it is closely tied to emotion and seems to be advantageous to survival in line with Darwinian natural selection[40, 101]. It is sometimes claimed that swimming is the best exercise one can do since it requires one to work nearly every group of muscles[40, 101]. The music can be thought of as the brain's analogue to swimming[40, 101]. It's a most basic and passive form; it exercises timing functions, matches patterns, and makes predictions[40, 101]. Like swimming as the best exercise to recover and retrieve body muscles, music has the ability to retrieve and remind the brain's internal information (Figure 6.3).

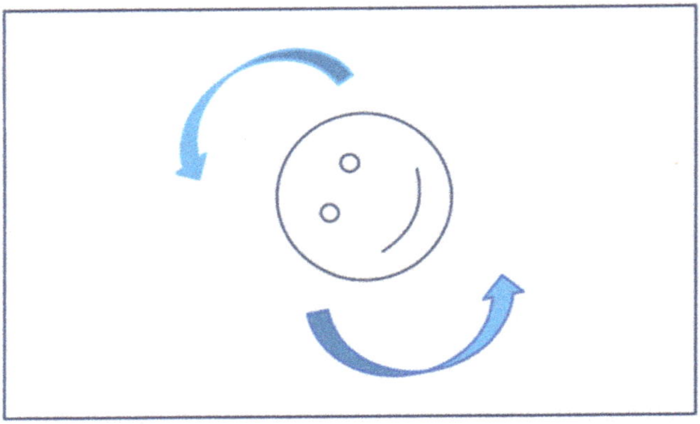

Figure 6.3 The music, a brain's analogue to swimming

It is necessary to design a technique for the music program in certain time periods. This technique forces the brain to manage fundamental memory to recall information without drug or surgical treatment. Based on the Meshk theory, a simple technique called 'special music program' is organized to restore memory and mind. This technique consists of two parts including harmonic music and repetitive periods. The synchronicity of body muscle movement is a basic principle in the swimming. Similarly, for a specific individual, musical harmony and repetition timing in this technique are key factors that have been overlooked in previous research on mind music therapy.

These important principles can be called spatial and temporal equivalence (for a particular individual). The repetition of harmonic music at specific intervals (certain periods) is a turning point in stimulating the mind to recall information in Alzheimer's disease. In addition, this novel technique is effective for restoring the human mind in other degenerative brain diseases as well as the consequences of an epidemic disease such as the covid-19 coronavirus. The synchronous neuron firing is like a symphony orchestra or philharmonic orchestra with many instrument. The harmony of playing music leads to create of a strong and impressive conclusion. Without each part of the symphony orchestra, the music is imperfect. The generation by synchronous firing is not the only important characteristic of conscious electromagnetic fields as in Pockett's original theory; spatial pattern is the defining feature of a conscious field[(37)]. In a philharmonic orchestra, there is an accurate spatial and temporal pattern that is well organized as it can lead to creating of a memorable artistic masterpiece. In Meshk's theory, spatial and temporal patterns are part of the intrinsic memory features. Now there is evidence that the cerebellum also stores the temporal information about the music we are listening to, and then it recalls this information while reproducing the music. Moreover, an amazing fact is that the cerebellum is discovered to be a center for emotions. According to Meshk Theory, the framework of the Mannea alphabet is interestingly consistent with the structural algorithm of the human brain. This alphabet could be the best basic format to receive and transmit data and information by the human. According to ancient literature, Sumerians had an alphabet with 37 letters, but the Mannea had an advanced alphabet that had 36 letters. Barbod is one of the first musician and composer in the ancient period of Kurdistan. He designed 360 types of music by 200 AC. Indeed, these 360 song are the basis of the music in Kurdistan. In the Sasanian period, music was popular and many musicians and composers worked on music. At that time, musicians were called Kheniagar (means to minstrel). According to the structure and function of the human brain, these 360 songs were created. Music is a part of mathematics and every song is a part of it. In mathematics, a circle has 360 degrees, and so these 360 songs correspond to it. The human memory is working in 360 degree, with two directions, and compatible with an entire circle.

The special technique of the music program is divided into 72 sections (36 pairs), each section having a specific song and time (Table 6.1). Each pair can be done in one day (one in the morning and one in the evening). Each period is 40 days, because the day after the 9 days of the program, there is 1 day to rest the mind. Therefore, there are nine repetition periods per year. In general, the program includes 324 working days and 36 rest days per year. This simple technique depends on the architecture and functions of human memory and thus is remarkable to retrieve memory in Alzheimer patients. Also, it is applied to prevent the disease for people in every age. The Mozart and Beethoven symphonies can be applied in this way, because these symphonies are compatible with the structure and function of human memory. The Kurdish and Iranian traditional music that was named Dastgah, including Mahur, Homayun, Nava, Segah, Cahargah, Rast-Panjgah, and Sur, can be very appealing. According to the Meshk theory, this adaptation is logical and it is not casual. The influence of music and symphonies on the mind has been investigated in the past decades, but is not organized in repeating certain courses (for a specific individual). That is why previous investigations on the music therapy of the mind have not succeeded and have remained as a cryptic problem up to now. The important points of this study can be used in mind music therapy research and other investigations. An Alzheimer's patient recover when expose to 36 harmonic musical melodies in a repetitive process. This action is repeated every 36 days and is performed 9 times during the year. These 36 pieces of music are divided into four sections as each partition contains 9 pieces of music. The spatial and temporal coordination of repetition is very important in this process because the process requires exact repetition and practice. There is a 4 day break between periods as well as a 5 day refresher to start the process again, so these 5 days can be a turning point. These periods of repetition and the interval between them stimulate the nerve cells to start normal activities. This technique can help memory re-manage the coupling process and thus remember brain information. In each period, part of the memory is retrieved and at the end the whole memory circuit is recovered. The function of this process is to stimulate the point 5 because the section is the organizing center of the memory. The coupling process is very crucial to re-connection of central and fundamental memory. Any disturbance in this process causes memory loss and eventually brain damage .As the scientist already proposed the combination of MT with other cognitive stimulations including dance and physical exercise can have more effective on the human memory[53]. Although music can simultaneously stimulate three parts of memory, including hearing, visual, and sensing, combining music with other cognitive stimuli such as dance and color can have a dramatic effect on recalling information (Figure 6.4). In Figure 6.4, the red, green, and blue circles represent the three main parts of memory and brain that are affected by music. The blue circle indicates the Water Equivalence (WE) driver and hearing system, and green circle indicates the Nunchaku Resonance (NR) and visual system. Also, the red indicates the Substance Proportion (SP) and the sensing system. Dance and color have been part of Kurdish culture since ancient times because they are important cognitive stimuli that can easily stimulate human memory. In Kurdish culture, colorful clothes are common that psychologically affect the happiness of society and increase life expectancy.

Number of Days	Timing		Number of Days	Partitions	Periods in the Year
	Evening	Morning			
40	Fifteen Minutes or More	Fifteen Minutes or More	4 × 9 = 36	M1	P1
			4 × 1 = 4	R1	
40	✓	✓	36	M2	P2
			4	R2	
40	✓	✓	36	M3	P3
			4	R3	
40	✓	✓	36	M4	P4
			4	R4	
40	✓	✓	36	M5	P5
			4	R5	
40	✓	✓	36	M6	P6
			4	R6	
40	✓	✓	36	M7	P7
			4	R7	
40	✓	✓	36	M8	98
			4	R8	
40	✓	✓	36	M9	P9
			4	R9	
360	✓	✓	(36 × 9) + (4 × 9)	M1 – M9 R1 – R9	**P1 – P9**

Table 6.1 The annual cycle pattern of music therapy timing in the Meshk technique

The sound of water and birds is a kind of harmonic music, with the color and smell of flowers, as well as the dance of plants are natural stimuli. Kurdish group dance, which is a kind of circular dance, with a single or a couple of figure dancers often added to the geometrical center of the dancing circle, is compatible with the geometrical form of human memory and brain. Therefore, this type of circular dance has an amazing effect on restoring the human mind and achieving balance. The combination of colors of Kurdish clothes with Kurdish dance is eye-catching and exciting. The Kurdish harmonic music and dance combined with wearing colorful clothes has an important effect on human intelligence and memory because it stimulates three parts of peripheral memory (visual, auditory and sense). That is why dementia and brain degenerative, especially Alzheimer's, are less common in Kurdistan than in other parts of the world. The Kurdish people wear a variety of colorful clothes as a symbol of the multi-colored world that stimulates the mind. Red, green and blue (RGB) is the main color in nature that is used to paint objects such as sash windows or to paint manuscripts and books in Kurdistan, as well as the color of Kurdish clothes. A sash window or hung sash window is made of one or more movable panels, or 'sashes', that form a frame to hold panes of glass, which are often separated from other panes by glazing bars, also known as Muntins in the US. In the Muntins' combination of colures in the Sash window is shown. It is likely that the Sash window art created for the first time in the Kurdistan because Sanadag city in the east of Kurdistan in known to Sash window city. Also, the city is known as one of the world's music cities, especially in playing the daf. Daf is a large Kurdish frame drum used in popular and classical music(89). Moreover, is it also used in religious ceremonies among Kurds(90).

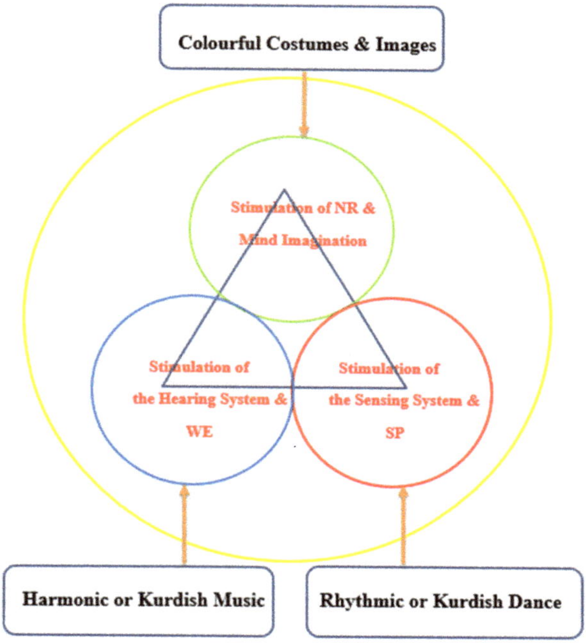

Figure 6.4 The effect of music, dance and color on the human memory cycle, the larger circle is the symbol of the brain, and the smaller circle is the type of effect on the memory

The row of memory unites in the memory strand is like to the row of nucleotides in Deoxyribonucleic Acid (DNA) as two snakes that wind together (Figure 6.5). Human Memory Strand (HMS) is a super structure and far superior than the DNA because many of the human intelligence functions such as imagination, induction, inspiration, meditation and so on that are not reasonable by the genetic science can be described by the memory strand. Therefore, the gigantic model of this structure is called 'Spiral Plate' which controls all the activities of the human body and intelligence. The snake is symbol of healthy and when two snakes hear a harmonic music wind together and make a shape like to structure of DNA. In Chinese culture, the dragon is considered a very important symbol. Li in chines means 'Lucid' and 'transparent'. The Lijiang River is like a sparkling, sinuous jade band, there being shallows at every band[28]. The music can clearly provoke the peripheral memory including visual, auditory and heart sense and therefore lead to recalling of information.

Figure 6.5. A schematic of double helix of memory strand similar to the Deoxyribonucleic Acid (DNA)

When the mind is affected by music two time in each day – in the morning and evening (or early night) continually, it causes to creation of the mind wormhole that can accelerate recalling of information in the memory. In other words, in this case, the connection between two strands of HMS is easily possible. In the book of Miracle, Morning by Hal Elrod[49] indicated that two hour of the early morning is miracle of the day. This is true because in the early morning brain is in its best mode. Furthermore, early night has also powerful effect on the brain and could has complement role for the early morning. Therefore, hearing the harmonic music in the both early morning and early night can have perfect impact on the human memory recalling (Figure 6.6). The effect of listening to harmonic music in the early morning and evening can be explained by the electromagnetic field of the brain. In modern physics, the electromagnetic field is understood to be not a classical field, but rather a quantum field; it is represented not as a vector of three numbers at each point, but as a vector of three quantum operators at each point. The most accurate modern description of the electromagnetic interaction (and much else) is quantum electrodynamics (QED), which is incorporated into a more complete theory known as the Standard Model of particle physics[82]. From outside, the ideal magnetic dipole is identical to that of an ideal electric dipole of the same strength[83].

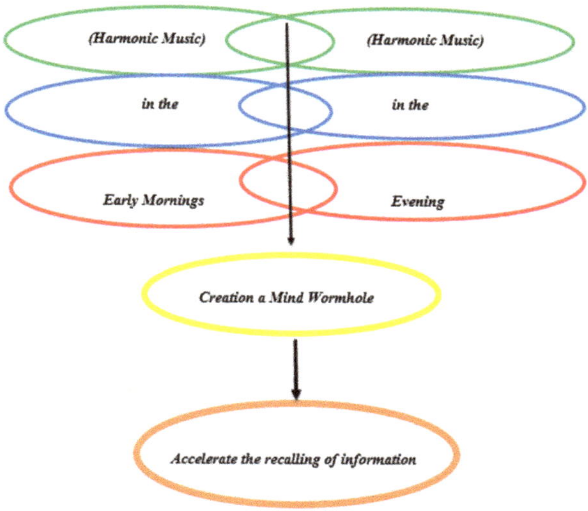

Figure 6.6. The comprehensive impact of music (in both morning and evening) on the human memory

The latest pandemic, Coronavirus COVID-19, has recently spread worldwide, leading to high mortality, especially in patients with Alzheimer's disease and other dementias. The Wall Street Journal reports that over the course of the past four months, while the world is under the thumb of the novel coronavirus pandemic, we've seen a surge in Alzheimer's deaths and deaths from related dementias, in which 15,000 more Americans have died from dementias than in the same timeframe in past years – an estimated 100,000 total between February through May(93). So far in 2020, the dementia and Alzheimer's fatality rate is nearly 20% higher than average from recent years(93). Some studies suggest that Alzheimer's disease could be a significant risk factor for more severe cases of COVID what the study showed was that the three biggest risk factors for death from COVID, in their hospitalized patients, were age over 70, lung disease, or Alzheimer's disease specifically(94). Alzheimer's disease specifically, independent of age, is a risk factor for severe COVID and possibly death from COVID(94). In addition, covid-19 can lead to dementia and brain degenerative disorders, especially Alzheimer's in the long term. As research into covid-19 continues, mounting evidence suggests that the disease may cause significant brain damage in some recovered patients(94). Some researchers have focused their attention toward the long-term effects of COVID-19 on brain health and the potential for developing neuropsychiatric conditions(94). Covid-19 is likely to affect human peripheral memory because studies have shown that heart attack, blindness and deafness are the early results of covid-19 seen in some patients. Due to the characteristics of peripheral memory, the malfunction of this memory affects the coupling process and coding rhythm in the memory. Therefore, this condition destroys the memory chain and over time leads to Alzheimer's and other neurodegenerative diseases. This situation, like a piece of wood inserted into a wheel, disables the memory cycle.

Music can affect the electromagnetic field of the heart and brain and thus coordinate the activities of the heart and brain. A ten-year-old boy in Italy underwent spinal surgery with a tumor on his spinal cord while a molecular biologist was playing the piano at his operating table like real-life 'grey's anatomy' episode in November 2020. A surgical team monitored electrical brain activity during the four-hour operation. The music was played at a particular frequency recognized to have a therapeutic effect on the body. When the music stopped, they saw changes in brain activity. Also, viral video of Marta Cinta Gonzalez Saldana, a former ballerina with Alzheimer's, shows vital role of music in memory. She memorized musical notes and played a swollen melody from Tchaikovsky Swan Lake. The particular ability of music to connect with people living with even severe dementia is becoming well known. Former concert pianist and music teacher Paul Harvey recently landed a No. 1 with Four Notes – Paul's Tune, a song he improvised at the piano despite his memory loss. Auditory memory may be the last thing to leave us because it's one of the first things we develop, at around 18 weeks in the womb, says Grace Meadows, programme director at Music for Dementia. "They're some of the deepest neural pathways that get laid down," she says. Our first language, the babbling and cooing of babies and their carers, is a form of musical exchange. “That’s why it’s so deep and why we’re so responsive to it, because it’s so primal.”

Music can be repeated at two times of the day, each lasting fifteen minutes or more. This type of repetition has a great effect on the connection and alignment of the electromagnetic field of the heart and brain. This method leads to heart and brain health and can be used in the treatment of heart and brain disorders without wasting time, medical costs and human suffering. One of the important features of the Meshk technique is that this method can be used both individually and in groups. The results of present study inspire composers to create counterpart music or ‘Couple Music’. This type of music can be compatible with the structure and function of human memory. Also, along with a certain type of dance like Kurdish dance, it has a great effect on memory. In addition, the study generates an idea to create a new type of triathlon, swimming-horseback riding and archery, which have important effects on human memory. It is necessary to conduct extensive research on Kurdish music and dance to discover more unknown aspects of it.

The black holes might serve as gateways to parallel universes and that missing information might actually be transferred to those universes

Stephen Hawking

Chapter 7
Conclusion

- The proposed structure for human memory in the Meshk theory called Human Memory Strand (HMS), as a super-DNA structure with excellent regenerative ability, is a turning point in research into the prevention and treatment of brain degenerative diseases, especially Alzheimer's, because based on HMS, a simple and cost-free method called Meshk technique has been developed to restore human memory. Therefore, the technique can be used effectively for healthy people to prevent Alzheimer's disease as well as people with Alzheimer's at any stage of the disease to retrieve memory. This practical method is based on a simple process that involves extracting information from the fundamental memory, as a vast repository of information, by stimulating all three parts of the peripheral memory including visual, auditory and sense using music over special time periods and regular repetitions. The structural details and time scheduling of the technique are also well described in the theory.

- In Alzheimer's and Parkinson's disease, peripheral memory is impaired. Impairment of regular stimulation of the sense part or hearing and visual sectors will upset the balance of peripheral memory. when the peripheral memory does not work properly in Alzheimer's, the connection to the fundamental memory will be lost over time, in which case the memory strand will be disintegrated. In Parkinson's disease, this connection appears to be frequently established in some parts of the memory strand and broken in some places. The main symptoms of this disease are caused by the frequent disconnection between peripheral and fundamental memory, which show itself in the form of tremor. The solution to treating Alzheimer's disease lies in the peripheral memory, so that the type, amount and timing of information that a person puts in their peripheral memory will determine the extent of progress in the treatment of this disease. Just like a water or wind turbine as which the amount of water or wind flowing into it will determine the final amount of electricity produced, the quality and quantity of information entering the brain that forms the peripheral memory will determine the amount and quality of information retrieval in the brain. In fact, the quality and quantity of information entered into the peripheral memory determines the amount and quality of information retrieved from the fundamental memory. Regular and well-planned entry of information for each specific person causes the brain gears to restart. In this way, the regular movement of the peripheral memory gears causes the central memory gears to move and eventually activates the fundamental memory gears.

- The Meshk technique is designed to stimulate both the right and left sides of the brain. In fact, the continuous and regular daily activity of the brain in the morning and evening with a specific and repeatable schedule plays the role of triggering both the right and left parts of the human brain. The repetition of a regular program during every nine days as a regular daily rotational program is the driving force of the human brain and activates memory in a regular rhythm. This is like the simultaneous transfer of power from the car engine to the front and rear wheels of the car and causes all four wheels to move in coordination with together so that the lack of movement of the front or rear wheels causes the car to stop. This rotational movement in the brain integrates the structure of the brain and memory and establishes the connections of neurons in the throughout the human brain especially between the right and left parts.
- The repetition of harmonic music in the special time periods is a turning point in stimulating of peripheral memory for recalling information in the Alzheimer disease. In the Meshk technique, nine repetition periods per year are considered, so that

each period consists of 40 days and is performed twice a day, in the morning and in the evening. Also, each course itself is divided into four parts, in each part there are 9 program days and one day off. Therefore, there are 36 running days and 4 closing days in each period of the program. Due to the structural features of the memory strand in humans, this type of music repetition shows the greatest fit and compatibility with human memory and therefore has a great impact on it.

A great question that no one has ever asked but is answered in this book based on the Meshk theory is as follows:

- ***Question:*** Why are the brain degenerative diseases, especially Alzheimer's, so rare in Kurdish areas unlike cancers?
- ***Short answer:*** Dance and music are an important part of the daily life of the people in Kurdistan and also their cultural symbol from ancient times until now. According to archaeological evidence, Kurdish music and dance are among the most ancient dances and music in the world. The Kurdish dance, as a beautiful group dance, is based on the structural pattern of human memory and brain. The originality of Kurdish music and dance has been preserved for thousands of years. Therefore, the combination of Kurdish music and dance, along with wearing a variety of colorful clothes on a regular basis, especially by women, simultaneously

stimulate all three parts of the peripheral memory, including visual, hearing and sense. This is the main cause of the lack of neurodegenerative diseases in this region.

- The Meshk technique shows a combination of music, dance and color, including Kurdish music-dance, as a triple Meshk technique that has the highest impact on all three parts of peripheral memory. It is a group dance that symbolizes the harmony, unity, equivalence and stability of the universe. The originality of Kurdish music-dance has been preserved for thousands of years. Finally, given that this book shows that Kurdish music and dance have amazing effects on the human memory and brain, it is necessary to study the Kurdish dance and music in advanced brain research as well as neurodegenerative disease especially Alzheimer's.

References

1. Adolphs, R., (2015). The unsolved problems of neuroscience. Trends Cogn. Sci. 19, 173–175. DOI: 10.1016/j.tics.2015.01.007.

2. Andersen, P., (1990). Synaptic integration in hippocampal CA1 pyramids. Prog. Brain Res. 83, 215–222. DOI: 10.1016/S00796123(08)612510.

3. Baddely, A., (2007). Working Memory: Theories, Models, and Controversies. Annual Review of Psychology: Volume 63, 2012.

4. Bell, G., Gemmell, J., (2016). Total Recall: How the E-Memory Revolution Will Change Everything. Computer and Technology, programming: Amazon Publication.

5. Apostolova, LG., (2016). Alzheimer Disease, Lifelong Learning in Neurology; 22 (2, Dementia):419–434. DOI: 10.1212/CON.0000000000000307, PMID: 27042902.

6. Buzsáki, G., (2010). Neural syntax: cell assemblies, synapsembles and readers. Neuron 68, 362–385. DOI: 10.1016/j.neuron.2010.09.023.

7. Diakonof, I, M., (1985). The Cambridge History of Iran: Media, Cambridge University Press, Cambridge.

8. Eysenck, M.W., (2012). Learning, memory and forgetting. Book review: Amazon Publication.

9. Hamarash, S., (2013). "Who is the Kurd?" YPS-Publishing, London, UK.

10. Hebb, D. O., (1949). The Organization of Behavior: A Neuropsychological Approach. New York, NY: John Wiley & Sons.

11. HerculanoHouzel, S., (2009). The human brain in numbers: a linearly scaled up primate brain. Front. Hum. Neurosci. 3:31. DOI: 10.3389/neuro.09.031.2009.

12. Kudrimoti, H. S., Barnes, C. A., and McNaughton, B. L., (1999). Reactivation of hippocampal cell assemblies: effects of behavioral state, experience and EEG dynamics. J. Neurosci. 19, 4090–4101.

13. Li, M., Liu, J., and Tsien, J. Z., (2016). Theory of connectivity: nature and nurture of cell assemblies and cognitive computation: Frontiers in Neural Circuits10:34. DOI: 10.3389/fncir.2016.00034.

14. Miguel, A.L., (2006). Septal serotonin depletion in rats facilitates working memory in the radial arm maze and increases hippocampal high-frequency theta activity. European Journal of Pharmacology: Volume 734, 5 July 2014, Pages 105–113.

15. Nicolelis, M. A. L., Fanselow, E. E., and Ghazanfar, A. A., (1997). Hebb's dream: the resurgence of cell assemblies. Neuron 19, 219–221. DOI: 10.1016/s08966273(00)809320.

16. Tsien, J. Z., (2007). The memory code. Sci. Am. 297, 52–59. DOI: 10.1038/scientificamerican070752.

17. Tsien, J. Z., (2016). CreLox neurogenetics: 20 years of versatile applications in brain research and counting. Front. Genet. 7:19. doi: 10.3389/fgene.2016.00019.

18. Tsien, J. Z., Huerta, P. T., and Tonegawa, S., (1996). The essential role of hippocampal CA1 NMDA receptor-dependent synaptic plasticity in spatial memory. Cell 87, 1327–1338. DOI: 10.1016/s00928674(00)818279.

19. Wallace, D. J., and Kerr, J. N. D., (2010). Chasing the cell assembly. Curr. Opin. Neurobiol. 20, 296–305. DOI: 10.1016/j.conb.2010.05.003.

20. Xie K., Fox G.E., Liu J., Lyu C., Lee J.C., Kuang H., Jacobs S., Li M., Liu T., Song S., Tsien J.Z., (2016). Brain Computation Is Organized Via Power-of-Two-Bases Permutation Logic.

21. Zadok, R., (2002). The ethno-linguistic character of Northwestern Iran and Kurdistan in the Neo-Assyrian Period, Archaeological Center, Michigan, USA.

22. Mehrabian, Albert, (1971). Silent Messages (1st ed.). Belmont, CA: Wadsworth. ISBN 0-534-00910-7.

23. Seyedielmabad, M., (2020). Introducing a Novel Approach to Study the Construction and Function of Memory in Human beings – the Meshk Theory: New Frontiers in Brain-Computer Interfaces. IntechOpen Publication, London. DOI: 10.5772/intechopen.87991.

24. Medina, J., (2009). Brain rules: Principles for Surviving and Thriving at Work, Home and School. Pear Press, Perseus Book Group, ISBN: 978-0979777745.

25. Murphy, J., (2002). Think Yourself to Health, Wealth, & Happiness, Prentice Hall Press; 1st Edition, ISBN-13: 978-0735203631, 352 P.

26. Murphy, J., (1963). The Power of Subconscious Mind, Self-help Book, Penguin Group, 375 Hudson Street, New York, USA, NY 10014.

27. E. B. Banning., (2011). "So Fair a House: Göbekli Tepe and the Identification of Temples in the Pre-Pottery Neolithic of the Near East", Current Anthropology, 52.5, 619 ff.

28. Chin chen, H., (2005). 100 wonders of china, Greenland International Book Co. Ltd. Rebo International b.v., Lisse, The Netherland, ISBN 978 90 366 1660 7.

29. Hawking, S. W., (1974). "Black hole explosions?" Nature. 248 (5443): 30–31. Bibcode: 1974, Natur. 248...30H. DOI: 10.1038/248030a0. ISSN 1476-4687.

30. Edwards, Owen, (2010). "The Skeletons of Shanidar Cave". Smithsonian. Retrieved 17 October 2014.

31. Murray, Tim., (2007). Milestones in Archaeology: A Chronological Encyclopedia. ABC-CLIO. p. 454. ISBN 9781576071861.

32. Wang F, Schoenbaum G, Kahnt Thorsten., (2020). Interaction between human orbitofrontal cortex and hippocampus support model based inference. PLOS Biology. Neural mechanisms underlying model-based inference.

33. Gonzalez-Carter, D., Fen Leo, B., Ruenraroengsak, P., Chen, S., Goode, A.E., Theodorou, I.G., Chung, K.F., Carzaniga, R., Shaffer, M.S.P., Dexter, D.T., Ryan, M.P., and Porter A.E., (2017). Silver nanoparticles reduce brain inflammation and related neurotoxicity through induction of H2S-synthesizing enzymes. Scientific Reports of Nature Research. DOI: 1038/srep42871.

34. NASA (2013). Near Earth Object Program FAQ, "What Are Asteroids and Comets", retrieved 30 July 2013.

35. BBC Two., (2003). Homeopathy: The test. Transcript, Retrieved 4 March 2007.

36. Umezawa, H., (1993). Advanced Field Theory: Micro, Macro and Thermal Physics. American Institute of Physics.

37. Pockett, S., (2000). The Nature of Consciousness. ISBN 978-0-595-12215-8.

38. McFadden, J., (2002). "The Conscious Electromagnetic Information (Cemi) Field Theory: The Hard Problem Made Easy?" Journal of Consciousness Studies. 9 (8): 45–60.

39. Fries, P.; et al. (1997). "Synchronization of Oscillatory Responses in Visual Cortex Correlates with Perception in Interocular Rivalry". PNAS. 94 (23): 2699–2704. Bibcode: 1997PNAS...9412699F, DOI:10.1073/pnas.94.23.12699. PMC 25091. PMID 9356513.

40. Kaliuk, A., (2012). Review on the book "Brain, Mind and Cognition, This is Your Brain on Music" by Daniel J. Levitin: The Science of a Human Obsession.

41. Jedlicka, P., (2017). Revisiting the Quantum Brain Hypothesis: Toward Quantum (Neuro) biology: Frontiers in Molecular Neuroscience. doi: 10.3389/fnmol.2017.00366, PMCID: PMC5681944.

42. Singer W., (2015). "Complexity as Substrate for Neuronal Computations", in Complexity and Analogy in Science: Theoretical, Methodological and Epistemological

Aspects, eds Arber W., Mittelstra J.ß, Sánchez Sorondo M., editors. (Vatican City: Pontifical Academy of Sciences), 209–218.

43. Markov N. T., Kennedy H., (2013). The importance of being hierarchical. Curr. Opin. Neurobiol. 23 187–194. 10.1016/j.conb.2012.12.008 [PubMed] [CrossRef].

44. London M., Häusser M., (2005). Dendritic computation. Annu. Rev. Neurosci. 28 503–532. 10.1146/annurev.neuro.28.061604.135703 [PubMed] [CrossRef].

45. Deco G., Jirsa V. K., (2012). Ongoing cortical activity at rest: criticality, multistability, and ghost attractors. J. Neurosci. 32 3366–3375. 10.1523/JNEUROSCI.2523-11.2012 [PubMed] [CrossRef].

46. Moseley, GL; Gallace, A; Spence, C., (2008), "Is mirror therapy all it is cracked up to be? Current evidence and future directions", PAIN 138 (1): 7–10, doi: 10.1016/j.pain.2008.06.026, PMID 18621484.

47. Ramachandran, V. S.; Rogers-Ramachandran, D. C., (1996), Synaesthesia in Phantom Limbs Induced with Mirrors, Proceedings of the Royal Society of London (263(1369)): 377–386, doi:10.1098/rspb.1996.0058, PMID 8637922.

48. Ramachandran, V. S.; Rogers-Ramachandran, D. C.; Cobb, S., (1995). "Touching the Phantom", Nature (377): 489–490.

49. Elrod, H., (2012). The Miracle Morning: The Not-So-Obvious Secret Guaranteed to Transform Your Life. 147 P.

50. Raz, N., Striem, E., PundaK, G., Orlv, T., Zohary, E., (2007). "Superior Serial Memory in the Blind: A Case of Cognitive Compensatory Adjustment". Publishing in Current Biology 17, 1–5, 3 July 2007. DOI 10.1016/j.cub.2007.05.060.

51. Pasqualotto, A., Proulx, M., (2013). Daily Mail Reporter, published: 01:30 BST, 29 April 2013.

52. Alberini, C., (2011). Spider-like Brain Cells May Be Active in Memory Making, Mount Sinai School of Medicine. PBS News Hour, Science.

53. Fang, R., Ye, S., Huangfu, J., Calimage, D.P., (2017). Music Therapy is a Potential Intervention for Cognition of Alzheimer's disease: a mini-review. Translational Neurodegeneration. DOI 10.1186/s40035-017-0073-9.

54. Zhang S, Zhang M, Cai F, Song W., (2013). Biological function of Presenilin and its role in AD pathogenesis. Transl Neurodegener, 2:15.

55. Hosseini SM, Kramer JH, Kesler SR., (2014). Neural correlates of cognitive intervention in persons at risk of

developing Alzheimer's disease. Frontiers Aging Neuroscience: 6:231.

56. Huang P, Fang R, Li BY, Chen SD., (2016). Exercise-related changes of networks in aging and mild cognitive impairment brain. Frontiers Aging Neuroscience, 8:47.

57. Hickman RA, Faustin A, Wisniewski T., (2016). Alzheimer disease and its growing epidemic: risk factors, biomarkers, and the urgent need for therapeutics. Neurol Clin, 34:941–53.

58. Mendiola-Precoma J, Berumen LC, Padilla K, Garcia-Alcocer G., (2016). Therapies for prevention and treatment of Alzheimer's disease. Biomed Res Int. 2016, 2589276.

59. Raglio A, Filippi S, Bellandi D, Stramba-Badiale M., (2014). Global music approach to persons with dementia: evidence and practice. Clin Interv Aging, 9: 1669–76.

60. Satoh M, Yuba T, Tabei K, Okubo Y, Kida H, Sakuma H, et al., (2015). Music therapy using singing training improves psychomotor speed in patients with Alzheimer's disease: a neuropsychological and fMRI study. Dement Geriatr Cogn Dis Extra, 5:296–308.

61. Herholz SC, Herholz RS, Herholz K., (2013). Non-pharmacological interventions and neuroplasticity in early stage Alzheimer's disease. Expert Rev Neurother, 13:1235–45.

62. Arroyo-Anlló EM, Díaz JP, Gil R., (2013). Familiar music as an enhancer of self-consciousness in patients with Alzheimer's disease. Biomed Res Int. 2013:752965.

63. Simmons-Stern NR, Budson AE, Ally BA, (2010). Music as a memory enhancer in patients with Alzheimer's disease. Neuropsychologia. 2010; 48:3164–7.

64. Palisson J, Roussel-Baclet C, Maillet D, Belin C, Ankri J, Narme P., (2015). Music enhances verbal episodic memory in Alzheimer's disease. J Clin Exp Neuropsychol, 37:503–17.

65. Gómez Gallego M, Gómez García J., (2015). Music therapy and Alzheimer's disease: cognitive, psychological, and behavioural effects. Neurologia, DOI: 10.1016/j.nrl.2015.12.003.23.

66. Sarkamo T, Tervaniemi M, Laitinen S, Numminen A, Kurki M, Johnson JK, et al., (2014). Cognitive, emotional, and social benefits of regular musical activities in early dementia: randomized controlled study. Gerontologist, 54:634–50.

67. Bruer RA, Spitznagel E, Cloninger CR., (2007). The temporal limits of cognitive change from music therapy in elderly persons with dementia or dementialike cognitive impairment: a randomized controlled trial. J Music Ther, 44:308–28. 26.

68. Ozdemir L, Akdemir N., (2009). Effects of multisensory stimulation on cognition, depression and anxiety levels of mildly-affected Alzheimer's patients. J Neurol Sci, 283:211–3. 27.

69. Kim HJ, Yang Y, Oh JG, Oh S, Choi H, Kim KH, et al., (2016). Effectiveness of a community-based multi-domain cognitive intervention program in patients with Alzheimer's disease. Geriatr Gerontol Int, 16:191–9.

70. Narme P, Clément S, Ehrlé N, Schiaratura L, Vachez S, Courtaigne B, et al., (2014). Efficacy of musical interventions in dementia: evidence from a randomized controlled trial. J Alzheimer's Dis, 38:359–69.

71. Sakamoto M, Ando H, Tsutou A., (2013). Comparing the effects of different individualized music interventions for elderly individuals with severe dementia. Int Psychogeriatr, 25:775–84.

72. Guetin S, Charras K, Berard A, Arbus C, Berthelon P, Blanc F, et al., (2013). An overview of the use of music therapy in the context of Alzheimer's disease: a report of a French expert group. Dementia (London), 12:619–34.

73. Simmons-Stern NR, Deason RG, Brandler BJ, Frustace BS, O'Connor MK, Ally BA, et al., (2012). Music-based memory enhancement in Alzheimer's disease: promise and limitations. Neuropsychologia, 50:3295–303.

74. BBC History., (2017). "Ancient History in depth: Mesopotamia", retrieved 21 July 2017.

75. Wilkinson, M., (2011). A Photographic History of World War II. Photographs by Daily Mail. Paragon Queen Street House, 4 Queen Street, Bath BAI IHE, UK. ISBN: 978-I-4454-3968-8.

76. Westwell, I., (2008). The Ultimate Illustrated History of World War I. [Illustrated with More Than 500 Photographs, Maps and Battle Plans. A Concise Authoritative Account the Course of the Great War, With Analysis Decisive Encounter and Landmark Engagement], Herms House.

77. Armitage, M., Lewin, L., Stanier, J., Charman, T.,. Kornicki, P., Pimlott, J., Tiedeman, G.T., (1990). Witness the 2175 Tumultuous, Tragic and Triumphant Days That Shook the World, World War II Day by Day. A Penguin Company, London, New York, Munich, Melbourne, Delhi.

78. Tracy, B., (2002). Focal Point: A Prevent System to Simplify your Life, double your Productivity, and Achieve all your Goals. AMACOM American Management Association. Management Association, 1601 Broadway, New York, NY 10019.

79. Super Brain Yoga Seminar, (2020). Yoga Bond Temple, Yoga – Meditation – Healing. Road 33, Pepper Street 3, Kep 15000, Cambodia.

80. Cox, D., (2018). What Stephen Hawking taught us about black holes, Turns out, black holes aren't so black. Mach, NBC News. NBCUniversal300 New Jersey Avenue, NW Suite 700 Washington, DC 20001.

81. Frohlich H., (1968). Long-range coherence and energy storage in biological systems. International Journal of Quantum Chemistry, 2(5): 641–649. DOI: 10.1002/qua.560020505. Bibcode: 196HJQC.2.641F.

82. Feynman, R., (2006). QED: the strange theory of light and matter. Princeton University Press. ISBN 978-0-691-12575-6.

83. Cullity B. D.; Graham C. D., (2008). Introduction to Magnetic Materials (2 ed.). Wiley-IEEE. p. 103. ISBN 978-0-471-47741-9.

84. BlackStock, E. G., (1978). Cerebral Asymmetry and the Development of Early Infantile Autism. Journal of Autism and Childhood Schizophrenia, 8,339–353.

85. Overy, K., Molnar-Szakacs, I., (2009). Being Together in Time: Musical Experience and the Mirror Neuron System: Musical Experience and the Mirror Neuron System, P 489–504.

86. Libet, Benjamin, (1981). "The Experimental Evidence for Subjective Referral of a Sensory Experience Backwards in Time: Reply to P.S. Churchland". Philosophy of Science. 48 (2): 182–197. DOI: 10.1086/288990.

87. Libet, Benjamin; Gleason, Curtis A.; Wright, Elwood W.; Pearl, Dennis K., (1983). "Time of Conscious Intention to Act in Relation to Onset of Cerebral Activity (Readiness-Potential) – The Unconscious Initiation of a Freely Voluntary Act". Brain. 106 (3): 623–642.

88. Nekoogar, Farzad, (1996). Traditional Iranian Martial Arts (Varzesh-e Pahlavani). pahlvani.com: Menlo Park. Accessed: 2007-02-08.

89. Tony Langlois, (2017). Non-Western Popular Music. Routledge. p. 91. ISBN 978-1351556156.

90. Iranica Online, (2020). "Daf and Dayera", retrieved 24 March 2020.

91. Waltari, M., (1945). Sinuhe the Egyptian. Historical Novel, Print (Hardback and Paperback), ISBN 1-55652-441-2 (English translation by Naomi Walford): 785 P.

92. Ahmed, Mohammed M. A., (2008). A Fire in My Heart: Kurdish Tales. World folklore series. Diane Edgecomb, Mohammed M. A. Ahmed, Çeto Özel. Libraries Unlimited. pp. 107–108. ISBN 9781591584377.

93. Marvar, A., (2020). Being patient, the latest news on Alzheimer's disease and brain health research, Five Reason the Coronavirus Pandemic Has Caused a Surge in Alzheimer's Deaths.

94. Troyer, E., (2020). Being patient, The latest news on Alzheimer's disease and brain health research, What Are the Long-Term Effects of COVID-19? Looking to the Future.

95. Lamb, R., (2020). How Houseflies Work, Housefly Anatomy: Wings, Legs and Abdomen. HowStuffWorks.

96. Karakecili, F., (2019).The Kurdish Project. Kurdish Dance Identity In Contemporary Turkey, Learn About Kurdish Dance, The Kurdish Project.

97. Bidlisi, S., (1597). Sharafnama, The Kurdish edition of Şerefname, History of Kurdistan.

98. Henry George Liddell, Robert Scott, A Greek-English Lexicon, on Perseus Digital Library.

99. Etymological Dictionary of Greek, (2009). The word meant "a kind of cake of roasted wheat-grains preserved in honey"; the Egyptian pyramids were named after its form (R. S. P. Beekes, Brill, p. 1261).

100. Leggieri, M., Thaut, M.H., Fornazzari, L., Schweizer. T.A., Barfett, J., Munoz, D.G., Fischer, C.E., (2019). Music Intervention Approaches for Alzheimer's Disease: A Review of the Literature: Frontiers in Neuroscience. DOI: 10.3389/fnins.2019.00132.

101. Kenealy S. In: Daniel J. Levitin, editor. Review on the Book "Brain, Mind and Cognition, this Is your Brain on Music". Dutton. 2012:314.

Statement of the Competing Financial Interest

The author does not have competing financial interests in the relation of the work and other affiliation. This is an individual study that is not affiliate to any institution or association. The author has investigated about the human memory, individually. This discovery is not affiliate to people or organization. The goal is providing a novel framework to investigate about the human memory. Progressing in this field could help to treat of the brain disorders especially the Alzheimer's.

CPSIA information can be obtained
at www.ICGtesting.com
Printed in the USA
LVHW012150090623
749359LV00009B/125

9 781685 622725